50 GREAT MYTHS OF HUMAN EVOLUTION

50 GREAT MYTHS OF HUMAN EVOLUTION

UNDERSTANDING MISCONCEPTIONS ABOUT OUR ORIGINS

John H. Relethford

WILEY Blackwell

This edition first published 2017
© 2017 John Wiley & Sons, Inc

Registered Office
John Wiley & Sons, Ltd, The Atrium, Southern Gate, Chichester,
West Sussex, PO19 8SQ, UK

Editorial Offices
350 Main Street, Malden, MA 02148-5020, USA
9600 Garsington Road, Oxford, OX4 2DQ, UK
The Atrium, Southern Gate, Chichester, West Sussex, PO19 8SQ, UK

For details of our global editorial offices, for customer services, and for information about how to apply for permission to reuse the copyright material in this book please see our website at www.wiley.com/wiley-blackwell.

The right of John H. Relethford to be identified as the author of this work has been asserted in accordance with the UK Copyright, Designs and Patents Act 1988.

All rights reserved. No part of this publication may be reproduced, stored in a retrieval system, or transmitted, in any form or by any means, electronic, mechanical, photocopying, recording or otherwise, except as permitted by the UK Copyright, Designs and Patents Act 1988, without the prior permission of the publisher.

Wiley also publishes its books in a variety of electronic formats. Some content that appears in print may not be available in electronic books.

Designations used by companies to distinguish their products are often claimed as trademarks. All brand names and product names used in this book are trade names, service marks, trademarks or registered trademarks of their respective owners. The publisher is not associated with any product or vendor mentioned in this book.

Limit of Liability/Disclaimer of Warranty: While the publisher and author have used their best efforts in preparing this book, they make no representations or warranties with respect to the accuracy or completeness of the contents of this book and specifically disclaim any implied warranties of merchantability or fitness for a particular purpose. It is sold on the understanding that the publisher is not engaged in rendering professional services and neither the publisher nor the author shall be liable for damages arising herefrom. If professional advice or other expert assistance is required, the services of a competent professional should be sought.

Library of Congress Cataloging-in-Publication data applied for

9780470673911 (Hardback)
9780470673928 (Paperback)

A catalogue record for this book is available from the British Library.

Cover image: Wlad74/Getty images

Set in 10/12.5pt Sabon by SPi Global, Pondicherry, India

Printed and bound in Malaysia by Vivar Printing Sdn Bhd

10 9 8 7 6 5 4 3 2 1

To the next generation:
My sons and my daughters-in-law,
Benjamin, David, Zane, Krissy, and Rachel,
and my nieces and nephews,
Adam, Burton, Dana, Evan, Katie, Maya, Melinda,
Noah, and Rebecca
L'dor v'dor

CONTENTS

Preface		x
Introduction: Myths and Misconceptions		1
1	**Ideas about Evolution**	**7**
#1	Evolution is a theory, not a fact	7
#2	Evolution is completely random	12
#3	All evolutionary changes are adaptive	16
#4	In evolution, bigger is always better	22
#5	Natural selection always works	26
#6	Some species are more evolved than are others	29
#7	Humans lived at the same time as the dinosaurs	33
2	**Human Origins**	**39**
#8	If apes evolved into humans, then apes should not exist today	39
#9	"*Ramapithecus*" was a human ancestor	44
#10	Humans and African apes split from each other over 15 million years ago	51
#11	*Gigantopithecus* was the ancestor of "Bigfoot" (assuming Bigfoot exists)	56
#12	Human traits all evolved at the same time	60
#13	Large brains evolved very early in human evolution	66
#14	The common ancestor of African apes and humans walked like a chimpanzee	72
#15	Bipedalism first evolved on the African grasslands	76

#16	Lucy was so small because she was a child	80
#17	*Australopithecus* was a killer ape	85
#18	Human evolution can be described as a "ladder"	90
#19	All hominin species have probably been discovered	97
#20	There are no transitional fossils in human evolution	101

3 Evolution of the Genus *Homo* — 109

#21	Only one species of *Homo* lived 2 million years ago	109
#22	Early *Homo* had modern human brain size	115
#23	Only humans are toolmakers and have culture	120
#24	We can identify species by the stone tools they made	125
#25	*Homo habilis* definitely made shelter	132
#26	Our ancestors have always made fire	135
#27	Early humans got all of their meat from hunting	139
#28	Species with larger brow ridges are more ape-like	143
#29	Neandertals walked bent over and were dumb brutes	148
#30	Neandertals definitely could not speak	153
#31	Modern humans appeared first in Eurasia	158
#32	"Mitochondrial Eve" is our only common female ancestor	163
#33	Neandertals did not interbreed with modern humans	169
#34	We do not need fossils any more to learn about human evolution	175
#35	All recent human species had large brains	178

4 Recent and Future Human Evolution — 187

#36	Each of us has billions of distinct ancestors	187
#37	The first Americans came from Europe or the Middle East	192
#38	The first Polynesians came from South America	198
#39	The origin of agriculture led to an improvement in health	203
#40	Civilization has been influenced by extraterrestrials	206
#41	The recent increase in life expectancy was due initially to antibiotics	211

#42	There are three distinct shades of human skin color	217
#43	Biological race is useful for understanding human variation	223
#44	All African Americans have the same genetic history	227
#45	Genetic ancestry is the same thing as cultural identity	231
#46	Sickle cell Anemia is a "black disease"	234
#47	There is a strong genetic relationship between brain size and intelligence test scores	239
#48	Humans are no longer evolving	243
#49	Blond hair will eventually disappear	246
#50	We can predict future human evolution	249

References   **258**
Index   **273**

PREFACE

I find that many people have a strong desire to learn about human evolution and our origins as part of a larger interest in the human condition. There are many ways to contemplate the origin and destiny of humanity, including the arts, literature, philosophy, religion, and science. A strong education in the liberal arts teaches us that there are many different ways to consider our nature and our place in the universe. This book deals with one aspect of the quest to understand the nature of humanity—using science to understand our existence as biological and cultural organisms subject to the evolutionary forces that affect all living creatures.

This book is not meant to be a textbook or a technical monograph. It represents my attempt as a teacher (I am a college professor) to explain a complex subject in a relatively short amount of space (and with the goal that you will go beyond my brief introductions to read and research topics of interest in more depth). Myths, mistakes, and misconceptions provide the focus for a broader treatment of the concepts, methods, and evidence for the history of our species. Above all, the study of human evolution is the study of human history, in the broadest possible sense, and it applies to *all* of us. No matter what else separates us all, the origin and evolution of humans is a history that we all share.

For a long time, I was interested in writing a trade book on human evolution, but had a hard time getting started. I have written several books, but my previous works were either college textbooks or books that wound up going into more specialized areas. Although I enjoyed researching and writing such books, I still wanted to have something on a more general level covering a wide range of topics in human evolution. For a while, I thought about trying to put together a book on "The top X things you should know about human evolution," where X was usually some number between 10 and 20, but never got started. I needed a hook, or a push.

In January 2011, I got both a hook and a push. A long-time colleague and friend, Rosalie Robertson, who was then a senior editor at Wiley-Blackwell, approached me and asked if I was interested in submitting a prospectus for a book on the "50 Great Myths of Human Evolution." After some thought, I realized that I could discuss many of the concepts of human evolution, past and present, within the structure of a book focusing on myths and misconceptions about human origins and evolution. After an extended delay due to a bout with cancer, this book is the result of that initial conversation. I am grateful to Rosalie for her vision and imagination and her patience with my questions and concerns. Thank you, my friend.

I have also benefited from the hard work and dedication of many people that worked on this project at Wiley-Blackwell. Thanks to Ben Thatcher and Mark Graney, who were involved in the initial submission and review process, and to Mark Calley and Tanya McMullin. Special thanks to the project editor, Roshna Mohan, for her patience with my endless questions and concerns, and to the copy-editor, Alta Bridges, for keeping track of endless details and for making the text more readable. I thank those colleagues that reviewed the initial proposal: David Begun (University of Toronto), Robin Dunbar (University of Oxford), Paul Lurquin (Washinton State University), Fred Smith (Illinois State University), Simon Underdown (Oxford Brookes University), and Bernard Wood (George Washington University). I am very grateful to Clark Larsen (Ohio State University) for his reading of both the proposal and the entire manuscript as well as answering specific questions. I am also grateful to Deborah Bolnick (University of Texas) and P. Thomas Schoenemann (Indiana University) for their assistance.

Finally, I have to give thanks to my wife, Hollie Jaffe, for support and guidance throughout this book, my career, and my life.

INTRODUCTION

Myths and misconceptions (or how and why I wrote this book)

What is a myth, and what are the myths of human evolution? I started giving these questions some thought several years ago when I was approached by the publisher to submit a proposal for a book to be called "50 Great Myths of Human Evolution." They had already published a book on misconceptions in psychology entitled *50 Great Myths of Popular Psychology* and were interested in publishing more books along the same line, focusing on the "50 great myths" of various fields, including human evolution. I was intrigued by the suggestion, as I had been contemplating a general book on human origins and evolution for a while. However, I was a little apprehensive about how to approach the idea of "myths" in human evolution. Like many words, the term "myth" has both narrow and broad meanings. My apprehension stemmed from a narrow interpretation of myth.

A narrow meaning of myth refers to the stories about Greek and Roman gods that I studied in a college mythology class. According to the online version of the *Oxford English Dictionary* (www.oed.com), a definition that would fit here is "A traditional story, typically involving supernatural beings or forces, which embodies and provides an explanation, etiology, or justification for something such as the early history of a society, a religious belief or ritual, or a natural phenomenon." This definition does not fit with what I wanted to do with this book as I wanted to go beyond the idea of simply examining stories about human origins, but instead wanted to look at different ideas and misconceptions

50 Great Myths of Human Evolution: Understanding Misconceptions about Our Origins, First Edition. John H. Relethford.
© 2017 John Wiley & Sons, Inc. Published 2017 by John Wiley & Sons, Inc.

regarding human evolution and, in particular, illustrate how scientific research often leads us to reject old ideas and consider new ones.

In this book, I use a broader definition of "myth" that is closer to the second definition given in the *Oxford English Dictionary*: "A widespread but untrue or erroneous story or belief; a widely held misconception; a misrepresentation of the truth." The myths in this book examine a number of ideas concerning human origins and evolution that fit this broader definition focusing on misconceptions—hence the subtitle of this book.

There are different types of misconceptions that exist when discussing human evolution. Some of these misconceptions are simply not true, but persist over time, such as the popular notion that much of our species' evolution was influenced by extraterrestrials (Myth 40). Some misconceptions arise from inaccuracies, incomplete data, and/or faulty assumptions, but somehow continue to perpetuate over time. An example from human evolution is the notion that the initial development of agriculture resulted in improved health (Myth 39), an idea possibly resulting from the faulty assumption that technological change in our species' evolution always results in progress across the board. Another common misconception is the idea that we no longer evolve (Myth 48), a conclusion reached only if we assume that our rapid cultural change completely negates biological change.

Many of the myths deal with topics that are not misconceptions at present, but refer instead to ideas that had once been considered accurate, but were later overturned because of new evidence and insights. Examples here include the idea that our early ancestor *Australopithecus* was a "killer ape" (Myth 17) and the notion that Neandertals walked bent over (Myth 29). Other myths look at ideas that have been questioned in recent times, but still remain on the table as possible hypotheses, such as the existence of only one species of the genus *Homo* two million years ago (Myth 21). These ideas are not "myths" in the classic narrow sense, but instead reflect shifts in consensus. Keep in mind that such shifts could in the future change further as new data become available. Ideas change as hypotheses are tested, and so might our conclusions on various myths and misconceptions. Today's "myth" might be tomorrow's consensus. It all depends on the evidence and the application of the scientific method.

The dynamic nature of science

Science means different things to people. Sometimes we narrowly equate "science" and "technology" such that recent developments in science often consist of lists of new inventions, drugs, and other important discoveries.

This is unfortunate because this narrow definition leaves out many interesting scientific discoveries (particularly those in human evolution) that have no direct or immediate practical benefit, but do inform us about the world and universe that we live in. It is also an unfortunate correspondence because, although science informs technology, that is not its only function or its essential nature.

At its core, science is a way of knowing, specifically a way of knowing about the natural world (including human behavior, as dealt with by the social and behavioral sciences). Although we sometimes think of science in terms of its direct benefits or the total accumulation of knowledge, it is most importantly a *process* that enables us to learn more about the physical world. Aspects of the scientific method will be described in a later myth, but, for the moment, we can break it down into a process of making observations, developing possible explanations for what we see (hypotheses), and testing them in some manner. Scientific evidence changes over time because this is a dynamic process as we ultimately discard hypotheses that have been rejected. In the general sense, a hypothesis is simply a proposed explanation. Some hypotheses can be supernatural (literally, "above nature") and invoke forces that we cannot directly perceive. To be a *scientific* hypothesis, we have to propose an explanation that is rooted in *natural* processes and is subject to testing.

A key feature of the scientific method is openness to being shown wrong. This does not mean that we *like* to have our hypotheses rejected (we don't) or that we don't resist new ideas and interpretations (we do). It means that ultimately we are open to sufficient evidence showing us that we were wrong, and that there might be a better way to look at things (although we might disagree with what is considered *sufficient*). In the jargon of the scientific method, we do not *prove* hypotheses so much as we fail to reject them (sort of like assuming someone is innocent until proven guilty). When a hypothesis is rejected in science, we throw it out and move on, coming up with a new explanation or modifying an old one. This is not always easy to do, as we are all subject to biases and feelings about pet hypotheses, but ultimately we reject or modify rejected hypotheses (or, if we do not, someone else will!).

This is a radical way of thinking about the world. Many times, we use a very different process of making decisions—we start with a conclusion and then pick data to support our established point of view. In an ideal sense, science works in an opposite manner, collecting all available evidence to test a hypothesis rather than assuming it is correct or incorrect beforehand. Of course, we are all human and are thus likely to be swayed by irrelevant information, wishful thinking, and preexisting biases.

However, as a process that is practiced by the scientific community, we can work through those sources of bias and error. We have to be willing to be wrong and say we are wrong. This is a difficult stance to make, because we often prize people for being resolute and standing for their convictions—admirable qualities but more appropriate to moral and ethical decisions than for scientific analyses. Imagine, for example, someone were running for elected office and made a statement about subject "X" that "I think that X is correct, but I remain open to the possibility that I am wrong." I am willing to bet money that this person would not be elected, as we often have little patience for people being on the fence or capable of "flip flops." In science, however, you *have* to be open to new evidence and ways of explaining them, *provided there is sufficient evidence*. As new evidence accumulates, ideas are repeatedly tested and often changed or thrown out. Some of yesterday's conclusions are now today's myths. This also means that some of today's conclusions might be tomorrow's myths!

In this context, I am always concerned at some of the reaction given to new scientific discoveries that appear to reverse previous ideas and conclusions, be they in human evolution, medicine, astronomy, or some other scientific field. Some people note these changes in a negative light, pointing out previous "errors" in judgment and analysis, and are left wondering why anyone would pay scientists that get things wrong. Well, the truth of the matter is that this is the way science is *supposed* to work. Our knowledge progresses by making hypotheses and testing them, and then throwing them out when they no longer fit the evidence.

It is in this spirit that I discuss the "myths" of human evolution. To be sure, many of these are completely settled (in my view), but others can change depending on new data and analysis. I try to be clear throughout about my views on current consensus as well as some additional possibilities. A warning, however, is that given the dynamic nature of science, it is quite likely that new evidence will shed further light on many of the topics covered in this book and will become out of date between the time I write these words and you read them. That is what is supposed to happen.

Structure of the book

I have picked 50 "myths" about human evolution that I find useful, particularly in teaching about human origins and evolution. (There are many more that could be discussed, but I accepted the number "50" to be part of the publisher's "50 Great Myths" series of books.) Each myth is designed to address a broader issue of science and of *paleoanthropology* (the study of human origins and evolution). I have broken the book into four sections. The first part examines some general myths and misconceptions about the nature of how evolution works. The second part focuses on human origins, examining the fossil record for the time between the initial divergence of African ape and human ancestors and the beginning of the genus *Homo*, including the evolution of bipedalism (upright walking). The third part continues looking at the fossil record in terms of the genus *Homo*, those species (including us) with larger brains, smaller faces, and reliance on a stone tool technology. The fourth and final part of the book examines recent (the last 12,000 years), current, and future human evolution, including the history of different human populations. Because evolution is a cumulative process, it is best understood in a linear manner from start to finish. Although I have tried where possible to make some myths independent, everything flows much easier if you read these myths in sequence.

If the idea of reading 50 essays seems daunting, remember that each myth is very short! The purpose of each essay is to use a myth or misconception to introduce a general topic in human evolution and provide some preliminary background and explanation. Each myth starts with a short "status" statement of several sentences that summarizes the thrust of the myth, also indicating if the topic has been settled or if there is still discussion on it. Because each myth is designed to be short, do not expect these to be complete reviews. The topic of every single myth can (and has) filled books. *The myths here are designed to be short introductions only.*

Although short essays have their purpose, you might find that you want more detail on the overall topic or on some of the specifics, or to read someone else's take on the issue. I have provided references to the facts and ideas discussed in each myth in a series of endnotes that are listed at the end of each section. A complete list of references is provided at the end of the book. Many of these references are to papers in academic journals that might not be available in many public libraries, but should be available at many colleges and universities. Some are also available for free on the Internet.

In general, I urge people interested in more detail on any of these subjects (or scientific subjects in general) to focus primarily on peer-reviewed journal papers and books. The peer-review process means that others in the field have examined the papers in terms of the soundness of the data, analyses, and arguments made. Peer review is a form of quality control and a researcher has to convince his or her peers that they have made the case for a particular conclusion. This is critical in modern times where anything and everything can be distributed on the Internet, often without any review. This does not make things on the Internet necessarily incorrect, but you have no guarantee of accuracy either. Peer review helps, as does looking at web pages that are connected to well-established scientific journals, magazines, and organizations (unless, of course, you subscribe to the notion that the scientific community consists of individuals involved in conspiracies, in which case I am not sure you will enjoy this book!).

1 IDEAS ABOUT EVOLUTION

In order to explore the myths of human evolution, we need to start with a brief review of how evolution works. It turns out that many of the myths of human evolution are related to misconceptions about the process of evolution in a general sense, starting with what is likely the biggest one of all—that evolution is "just a theory." This section of the book examines some common misconceptions of the process of evolution.

Myth #1 Evolution is a theory, not a fact

Status: This is a myth based on a misunderstanding about the use of the word "theory" in the natural sciences. When we state something is a theory, such as evolutionary theory, atomic theory, or the theory of gravitation, we are not suggesting that it may or may not exist (a more popular use of the word "theory"). Instead, we are talking about a hypothesis that has been tested repeatedly and has stood the test of time without being rejected.

Of all the myths about evolution, perhaps the one that we hear more than any other is the idea that evolution is a theory and not a fact. Most often, this myth is expressed as the statement "It's just a theory" or the somewhat longer "It's a theory, not a fact." By contrasting fact and theory, we are forced into an either-or situation. Either evolution is indeed a fact or it is a theory. We then must choose between one side and the other. According to popular logic, if we accept evolution as a theory then it is not necessarily a demonstrated fact. The logic works here only if we

50 Great Myths of Human Evolution: Understanding Misconceptions about Our Origins, First Edition. John H. Relethford.
© 2017 John Wiley & Sons, Inc. Published 2017 by John Wiley & Sons, Inc.

define the word "theory" as an unsupported or unproven hypothesis or explanation. In other words, if we classify evolution as "just a theory," it implies that evolution may or may not exist. In terms of human evolution (that aspect of evolution that tends to upset folks more than, say, elephant evolution because it is personal), the statement that evolution is "just a theory" means that humans may or may not have evolved. If we cannot tell, then evolution (including human evolution) is therefore not a fact. It is, according to this logic, at best an opinion.

Although much of the above may seem logical and perfectly reasonable, the argument rests on an underlying assumptions that "theory" means an untested hypothesis or mere opinion and that something can be either a fact or a theory. It turns out that our more popular use of the word "theory" is not what it means in the context of scientific thought. Evolution is actually both a fact and a theory. In my introductory course on biological anthropology, I ask the class on the first day to raise their hands if they think evolution is a fact. I then ask the class to raise their hands if they think evolution is a theory. I then tell them "Congratulations! All of you are correct. Evolution is *both* a fact and a theory." This statement can cause some consternation in anyone who is used to facts and theories being considered in terms of an either-or proposition. In order to see the mistake being made by this proposition, we need to consider a bit of the underlying philosophy and method of the natural sciences and explore briefly what we mean by fact, hypothesis, and theory.

To most of us, the definition of "fact" is pretty straightforward. A fact is a verifiable truth—something we can all observe and agree on. The key feature here is that facts must be capable of being verified. If I say that there are trees in my yard, you can actually look and see if this is true. Some facts are easy to verify and we will all agree with little or no argument. For example, if we drop an object, such as a pencil, it will drop to the ground. We call this fact gravity. Sometimes facts are contingent upon a more exact definition. In the case of gravity, the pencil would have to be dropped while standing on something of sufficient mass to generate sufficient gravitational force to attract the pencil. Sometimes facts are tricky because they are not directly observable with our senses. We can easily see a pencil dropping, but what of the fact that infectious diseases are caused by bacteria and viruses that are not visible to the human eye. Of course, we easily accept the existence of such microorganisms because we have developed microscopes and other technology to make our observations. However, imagine you were alive during the fourteenth century and someone explained to you that the Black Death (bubonic plague) was caused by a bacterium, something that could not be seen except with a microscope (that had not yet been invented). I suspect that most people at that time would have rejected this idea because the plague bacterium could not be observed with the naked eye.

Observing something, either directly with our senses or with technology, is a start in establishing a fact, but you need to remember that facts must be verified. Sometimes in the history of science, we find that our basic facts change when more observations are made. At one time, for example, it was thought that humans had 24 pairs of chromosomes, but over time, more advanced methods revealed that we actually had 23 pairs of chromosomes. At one time, a fossil known as Piltdown Man (discussed in Myth 13) was thought to be a fact supporting the then-popular view that humans evolved large brains before losing certain ape-like features of the teeth. In this case, inconsistency with other facts, development of better ways to date the individual fossils making up Piltdown Man, and other pieces of evidence pointed out that it was not a fact, but instead a fake. Someone (whose identity is still not known with certainty) faked the whole thing. Again, such lessons show us that science requires verification even with basic facts.

What about theory? Before considering the different meanings of the word "theory," we need to start with the idea of a hypothesis. Science is not simply an accumulation of facts about the physical universe. We also try to explain what we see. A hypothesis is just a tentative explanation of the facts. For example, why does a pencil fall to the floor when I let it go? In order to make my point about the nature of a hypothesis and how it ties into science, I am going to state an obviously ridiculous hypothesis to explain the falling pencil. Imagine that I have placed a magnet inside the pencil and then held it over a spot on the floor under which I have buried a very powerful magnet. When I let go of the pencil, the magnetic forces cause the pencil to drop to the floor. I imagine as you are reading this, you are thinking that this hypothesis is one of the silliest things you have ever heard, and so ridiculous that even discussing it is a complete waste of time. Yes, it is ridiculous and it is clearly false, but the interesting thing here is that my wacky idea is actually a good scientific hypothesis because it can be tested. There are a number of ways to test this hypothesis. Break open the pencil or dig under the floor to find there are no magnets. Use a device (such as a compass) and fail to detect any localized magnetic force. Or, in perhaps the most simple but also most elegant test, drop your own pencil (or shoe or baseball) and find that they all drop to the ground without any magnets being placed inside of them.

In each case, the hypothesis has been tested and has been rejected. We then have to move on to another hypothesis. Each time we develop a hypothesis we try to determine some way to test it. Science is continually involved with the testing and retesting of hypotheses, looking for hypotheses that have stood the test of time. In the natural sciences, we use the word "theory" to indicate a hypothesis, or set of hypotheses, that has been tested repeatedly and has not been rejected. We might continue to refine the theory, but the basic elements are widely agreed upon and unlikely to change.

This definition of theory contrasts with the popular idea that a theory is a hypothesis or just a guess and that the subject of the theory may or may not exist. However, when you hear the phrase "theory of gravity," do you think that gravity may or may not exist? Of course not. To take another example, consider atomic theory in chemistry. Does the inclusion of the word "theory" make you think, "Well, atoms are only a theory and they may or may not exist"? I doubt any reader takes this stand. The elements of atomic theory have been tested and have held up over time. The same is true for evolution. The basic ideas regarding the mechanisms of evolution (described in later myths) have been confirmed and form the basis for modern evolutionary theory. As with gravity and atoms, evolution is both a fact and a theory. Arguing that something has to be one or the other is a misuse of the scientific method.

Historically, we associate part of modern evolutionary theory with the insights of the nineteenth-century naturalist, Charles Darwin, who contributed to our understanding of both the fact of evolution and part of the underlying mechanism for evolutionary change. By Darwin's time, many in the scientific community were coming to grips with evidence showing changes due to evolution. The spread of the Industrial Revolution had led to increased mining and quarrying activity. As people dug into the earth, they found many fossils of creatures that did not fit nicely and neatly into their views on variation. Imagine, for example, you were digging in your backyard and found the skull of a cow. How would you explain it? Depending on where you live, the explanation might be very simple—perhaps your property was once a farm where cows lived and died. Or, imagine you unearthed a skull of a modern human. Although such a discovery might lead to all sorts of speculations about the identify and fate of the person you found, the simple truth is that finding a modern human skull in the ground is not likely to be an earth-shattering discovery.

However, what would you do if you found the remains of a creature that no longer lived, such as the bones of a dinosaur? This discovery implies that there were creatures that once existed but have since become extinct (which turns out to be quite common—we now know that over

99 percent of all species that have ever lived have become extinct). How do you explain this extinction? You then notice upon further examination that the bones of the creature you discovered are similar to, but not identical to, living creatures. For example, if you look at fossil remains from many millions of years ago, you will find creatures that are clearly similar to horses, but instead have three toes on each foot, as compared with the single toe typically found in modern horses. Or, in the case of human evolution, we can go back 2 million years ago in Africa and find creatures that are very similar to us in terms of how they walked and their basic body anatomy, but have smaller brains and larger faces. As we examine the fossil record even further, we see examples of trends over time, such as a reduction in the number of horse toes or the increase in the brain size of bipeds. Such trends are clear examples of evolution (and more will be presented throughout this book). How do you explain such facts?

Darwin was one of those who sought an explanation for change over time. Darwin made two very important contributions. First, he collected data confirming the fact of evolution as revealed from field studies of living organisms, the fossil record, and the comparative anatomy of different species, among other sources of evidence. His result was a convincing argument that all living species were related through a process of what he termed "descent with modification." The mechanism that Darwin proposed (natural selection) will be dealt with in later myths, but here we just focus on the fact that natural selection was a hypothesis relying on natural phenomena that explained the observed facts. As with all scientific hypotheses, Darwin's idea has been tested repeatedly. Because it has survived without refutation, the concept of natural selection has been elevated to the status of a scientific theory. Once more, keep in mind that the word "theory" has a very specific meaning here and does not mean something that may or may not exist.

The final point about Darwin's idea is that even though it forms part of modern evolutionary theory, his concept of natural selection is not the entire answer. Although Darwin got a lot right, he also had questions that remained unanswered during his life. The tentative nature of scientific explanation can be frustrating to those seeking a final definitive answer, but it is the basic nature of scientific inquiry with which we continue to refine our explanations. The theory of evolution is no exception. We do not have all the answers, but continue to seek them through the scientific process. However, although scientists continue to debate the details of the evolutionary process, there is agreement on both the fact of evolution as well as the basic explanation of how evolution happens. The details of the evolutionary process are described briefly in the next myth.

Myth #2: Evolution is completely random

Status: This is a myth because it implies that evolution is a chance event. Although some aspects of evolution (such as mutation) have a random element, other aspects, such as natural selection, are not random. Whether an individual survives and reproduces or not depends on their evolutionary fitness relative to their local environment. Like many natural processes, evolution has both nonrandom and random components.

A common misconception of the evolutionary process is that it is random; that is, due to chance. Taken to an extreme, this misconception can lead to a rejection of evolution altogether. After all, how could something as complex as the human body (or any other organism) be due to chance? That is analogous to scattering thousands of Scrabble™ tiles at random and having them spell out the Declaration of Independence. Complex sentences or biological structures, such as the human body, would seem to defy randomness, which many people equate with something "just happening." Part of the confusion may lie with the fact that some parts of the evolutionary process are random. However, having some randomness in parts of a process is not the same as an entire process being random. To be more specific, the origin of initial genetic variation is random, but the outcome is not. To see the distinction here, we need to look more closely at how evolution works.

As described in the last myth, Darwin's most significant contribution to the theory of evolution is the description of natural selection. Darwin noted that there is considerable biological variation in living creatures, something that we can all see easily. For example, not all birds look alike, but vary in terms of size, color, and other physical traits. As you walk down the street, you will see the same is true of humans; people vary in terms of size, shape, body proportions, skin color, hair color, and many other characteristics. This is even more apparent when looking beyond observable physical traits and we consider genetic traits where people vary in terms of blood types, blood proteins, and DNA markers, among others. Variation is all around us in the natural world; an observation that Darwin was able to tie to environmental differences.

Darwin also relied on the observation that more organisms are born than will survive to adulthood. For example, if a fish lays 100 eggs, it is a certainty that not all 100 offspring will survive to adulthood. Most will die, but some will survive. The same process of differential survival is true of all species—some individuals survive and reproduce, thus continuing the species, whereas others die before reaching reproductive age or fail to reproduce. Darwin tied together the observation of differential survival with the observation of variation. Given variation within a species, in a specific environment some individuals will be more likely to survive and reproduce than others. Imagine, for example, that there is variation in the size and shape of the beak of a bird in an environment where the main source of food is large seeds that are tough to crack open to eat. In such a case, those birds that have the most powerful beaks are most likely to eat and hence to survive. Consequently, the birds that are better adapted will contribute more to the next generation than those that are less adapted to the specific environment. Over time, the genetic characteristics of the population will change and large, powerful beaks will become more common.

The principles of natural selection are often best understood by analogy to the process of animal domestication. Imagine, for example, that you have just inherited a pig farm and you decide to go into the business of raising pigs for sale as food. When you first arrive on your new farm, you will notice that there is variation in the size of the pigs. Some of the pigs may be large and fat whereas others may be small and scrawny. Over time, you will sell off some pigs and keep others for breeding stock (because you want to produce additional generations of pigs). Keep in mind that you get a better price for the larger pigs. Which pigs do you sell and which pigs do you keep as breeders? If you are interested in long-term profitability you will ignore an impulse to sell the large pigs right away and instead you will keep them as breeders because of the common knowledge that, all other things being equal, larger pigs will produce larger offspring. This is not a perfect correlation, but it is strong enough that people have relied on this principle of selective breeding to feed themselves in the 12,000 years since agriculture has existed. The idea is simple enough to use even without knowledge of the underlying genetics—breed for the characteristic of interest and it will become more common over time, be it the size of pigs, speed of a horse, disposition of a dog, or many other traits. This selection is not random—the farmer does not roll dice or flip coins to pick which pigs are breeders.

Darwin recognized how this process of selection could lead to evolution, where the change over time was due to the farmer selecting who lived to reproduce and who did not. He also recognized that the same process could happen in nature, but where the selection was not the product of conscious manipulation by a human being, but was instead due to interaction with the environment. Those organisms that are better adapted to a given environment are more likely to survive and reproduce and will then pass on their characteristics in greater numbers to the next generation. Unlike the artificial selection that occurs due to the intervention of the farmer, this selection occurs in nature and is therefore termed natural selection.

A classic example of natural selection acting upon variation is found in studies of the coloration of the peppered moth in England. At one time, most of the moths of this species were light-colored, but a very small number were dark in color. The light color was more common because it was adaptive; the light color acted as camouflage when the moths rested on the light-colored tree trunks. Because these moths blended in, they were less likely to be seen by birds, unlike the dark-colored moths that were more visible and thus more likely to be eaten. Here, selection acted to maintain the light color over time and most dark-colored moths were selected out of the gene pool. Whether a moth was eaten or not was not random.

However, scientists also noted what happened when the environment changed because of industrial pollution killing off lichen on the trees, exposing the underlying dark color. At this point, the selective balance shifted and light-colored moths were then at a disadvantage and dark-colored moths were at an advantage. Each generation the proportion of dark-colored moths increased until they were the most common form as the population became better adapted to the environment.[1] Although this is a relatively small amount of change, the process of natural selection can apply to larger changes over geologic time, leading to major divergences.

Darwin's model of natural selection leaves out one important question—where does variation come from in the first place? Why are some pigs bigger than others? Why are some moths darker and some lighter? Darwin did not have the answer about the origin of variation; he noted its existence and then described how natural selection could act upon this variation, but lacked the insights of twentieth-century genetics that show us that the ultimate cause of genetic variation is the process of mutation.

A mutation is a random change in the genetic code, DNA. Mutations can occur for a number of reasons including the effect of background cosmic radiation, leading to an error in how the DNA is being copied. The DNA consists of sequences of four chemical bases and can be thought of as analogous to an alphabet with four letters that spells out the instructions that regulate all processes of life, ranging from the structure of proteins to the development of an organism. Some mutations involve a change in one of the letters (bases), while others can involve duplication or deletion of larger DNA sequences. Still other mutations involve movement of DNA sequences from one chromosome to another. Following the alphabet analogy, mutations act to change the message being transmitted.

Mutations can occur in any cell and interfere with biological function (such as leading to cancer). From an evolutionary perspective, we are interested in mutations that are transmitted through sex cells (sperm and egg in bisexually reproducing organisms).

Natural selection acts upon mutations. If a mutation is harmful to the organism that inherits it, hindering survival or reproduction, it can be eliminated through natural selection. Selection thus acts to weed out harmful effects. On the other hand, if a mutation leads to an advantage, it can be selected for and increase in frequency over time. Putting mutation and natural selection together, we get a picture of mutations generating variation that is then filtered by natural selection, leading to the reduction in frequency of harmful mutations and the increase in frequency of helpful mutations. (The actual picture can get much more complicated, but this view suffices for now.)

We can now turn to the question asked at the beginning of this myth—is evolution a random process? This question does not have a single yes or no answer. Mutation is a random process. Mutations do not appear when they are needed. (For example, a dark-color moth mutation did not appear in the moth population just because the environment changed.) Although we can measure the probability of a mutation occurring in any given organism in any given generation, we do not know for sure whether a specific DNA sequence will mutate or not at any given point in time. Think of the analogy of flipping a coin. If you are using a fair coin (no magic tricks allowed), you know that the coin will land heads up or tails up. For our purposes, the outcome is random. Although we do know that the probability of getting heads or tails is 50 : 50, we do not know beforehand whether any specific coin flip will be heads or tails. In terms of the moth example, whether a mutation leading to dark coloration appeared in a given generation or not is a random process. It is a matter of luck.

Does this mean that evolution is random and everything we see around us resulted merely from a series of chance events? Absolutely not. The fact that mutation is random simply means that the initial generation of variation is random, not the outcome. Remember, natural selection is not a random process. Whether an organism will survive and reproduce or not is a function of its adaptive value (what we call "fitness") in a given environment. When the trees in England became darker, the difference between survival of dark-colored and light-colored moths was not a matter of chance, but instead a direct outcome because of differences in fitness (because light-colored moths were more likely to be eaten). Although the direction of evolutionary change may change as the environment changes (as in the case of the peppered moth), this is not a random change. Although evolution does have a random component (mutation), the direction of evolutionary change due to natural selection is not a random outcome. Think of this difference in terms of how humans domesticated corn (or any other plant or animal). Humans altered the evolutionary course of corn to produce kernels that were large and stayed on the cob. They did this by the process of artificial selection acting upon the variation in corn that was available in nature. Although the initial origin of this variation was a random event due to mutation, the outcome of domesticated corn was not.

The discussion of how evolution works continues with the next myth. For the moment, it is important to discard ideas that the evolutionary process has to be entirely random or nonrandom. Evolution has both random and nonrandom (deterministic) components. It does not have to be just one or the other. To pursue an analogy with life, consider the movie *Forrest Gump*, where the title character muses about whether people have a destiny (deterministic) or whether we are "all just floating around accidental-like on a breeze" (random). Forrest wisely concludes, "Maybe both is [sic] happening at the same time."[2]

Myth #3 All evolutionary changes are adaptive

Status: This is a myth that results from equating the entire evolutionary process with natural selection acting upon mutations. Not all evolutionary changes reflect adaptation. There is also random fluctuation over time, known as genetic drift. Evolutionary biologists all agree that both selection and drift are important, although there is debate over the relative influence of each.

As described in the previous myth, natural selection is a powerful agent of evolutionary change, acting upon mutations to decrease the frequency of harmful mutations and increase the frequency of helpful mutations. Over time, species become better adapted to their environments, as seen in numerous field studies of living organisms. An example from the human species is the global distribution of skin color, where native populations at or near the equator tend to be the darkest, and populations farther away from the equator, north or south, are increasingly lighter. This pattern correlates with the global distribution of ultraviolet radiation. The story of skin color adaptation will be explored in detail in a later myth (Myth 42), but the point here is that variation in skin color can be explained by adaptation through natural selection to ultraviolet radiation.

The model of genetic variations introduced through mutation being acted upon by natural selection is both simple and elegant, and can explain the variation of many biological traits. Many examples from human evolution will be presented in later myths, including the origin of upright walking, the increase in brain size, and physical variations such as skin color. We also know that the long-term process of natural selection continues in recent times, as we have a number of examples of evolutionary changes in our species that have taken place within only the past 10,000 years or so.

It therefore may be tempting to explain all evolutionary change as the outcome of mutation and natural selection, and assess all physical and genetic changes in terms of their adaptive significance. Although this works for some traits, does it work for all? For example, some people have earlobes that are attached smoothly with the ear, and others have earlobes that are unattached and hang freely. Is there any adaptive significance to this variation? Could it possibly have anything to do with survival or reproduction? Should we postulate that some people are more attracted to potential mates depending on their earlobes? This seems a bit far-fetched to me. As another example, consider variation in different human blood groups (biochemical traits defined by reaction of surface molecules to various antibodies). There are a number of different blood group systems, with the ABO blood group and the Rhesus blood group being the best known because their biochemistry affects blood transfusion. (These are actually two distinct genetic systems, ABO and Rhesus, and when someone says their blood type is O negative, this is actually shorthand for type O blood for the ABO blood group system and Rh negative for the Rhesus blood group system.) There are many other blood group systems that are seldom typed for medical purposes, such as the MN, Diego, Duffy, and P blood groups, among others. All of these blood group systems show variation among human populations, and our job is to ask why these patterns exist. Can we explain differences in the frequencies of different blood group systems in terms of natural selection? For example, Native American populations typically have a higher frequency of blood type O for the ABO system than populations elsewhere in the world. Why? To take another example, the Basque populations in Europe have higher frequencies of Rh-negative type blood than elsewhere in the world. Does this reflect the past (or current) action of natural selection, or is something else going on?

There are many cases where natural selection provides the best explanation for patterns of biological variation. However, there are other cases where the evidence suggests that natural selection may not be the only factor contributing to variation. There are also cases where natural selection does not explain anything about the variations that we see. Can some variation be explained by nonadaptive evolutionary change? The answer is yes.

In the early part of the twentieth century, scientists grappled with the question of what causes evolutionary change. Laboratory and field studies were demonstrating the importance of both mutation and natural selection, but it became apparent that one could explain genetic change within a population (defined as a change in the frequency of different genetic variants) through the action of four mechanisms, termed evolutionary forces. Mutation is one of these four forces, and natural selection is another. A third evolutionary force is known as gene flow, which is the movement of genetic material from one population to another. Let's say that you leave your hometown and move somewhere else, marry someone who is living there, and then have a child with your spouse. Gene flow has occurred because you have mated with someone in a different population and thus have connected two populations genetically. Gene flow can affect the genetic makeup of a population in two ways. First, new genetic variants can be introduced into the population from somewhere else. This process allows new mutations to be spread throughout a species. Second, gene flow acts to make populations more similar to each other over time, much like mixing paint from two cans of paint, say red and white, will make the paint in each can eventually a similar shade of pink.

The fourth evolutionary force is the key one in our discussion of adaptive versus nonadaptive evolution and is known as genetic drift. The definition of genetic drift is short and to the point—random fluctuation in the frequency of a genetic variant over time—but a bit harder to conceptualize. As an example, let us consider a simple genetic trait in human populations, the MN blood group. The gene that controls the MN blood group has two different forms, or alleles. (Different forms of genes are known as alleles.) These two forms are the M allele and the N allele, and they correspond to different sets of instructions that produce different types of molecules on the surface of red blood cells. The M allele codes for type M molecules and the N allele codes for type N molecules. If you have inherited an M allele from both parents, you will have type M blood, and if you have inherited an N allele from both parents, you will have type N blood. If you inherit an M allele from one parent and an N allele from the other parent, you will have both types of molecules on the surface of your red blood cells and will therefore have type MN blood; this is because unlike some other genetic traits, neither the M nor the N allele is dominant.

With an example such as the MN blood group, we could go into a human population, take samples of everyone's blood, determine how many M and N alleles each person had, and count the number of M and N alleles in the population. Let us say that we do this and find that 55 percent of the alleles in the population are M alleles and 45 percent are N alleles. Now, imagine you return a generation later and find that the frequency of the M allele is now 58 percent. You have detected (at a very low level) evolution, defined here as a change in the allele frequency over time. The frequency of the M allele has changed from 55 percent to 58 percent. The trick is figuring out why this change occurred. Did the frequency of M increase over time because of selection, or did something else happen? Under genetic drift, it is possible for a frequency to change by chance. This is an example of nonadaptive evolution—there has been a change, but not due to natural selection.

As another example, imagine that you have gone to a population and measured everyone's head and found that the average length of a person's head was 180 millimeters. If you visit the population a generation later and find that the average head length is now 178 millimeters, what can you conclude? You might suspect natural selection, but we could also get changes of this magnitude due to random chance; that is, genetic drift.

Genetic drift is an example of what we call sampling error. To illustrate this, picture a large group of people where half have brown eyes and half have blue eyes. Now, imagine picking 10 people at random (in other words, not looking at their eye color). Although you might expect to get five people with brown eyes and five people with blue eyes, you also know that by chance you could get other outcomes, such as six people with brown eyes or three people with brown eyes, among other outcomes. (You can try this experiment by flipping coins, letting heads represent brown eyes and tails represent blue eyes.) Genetic drift works in a similar manner; this means that the frequency of an allele in the offspring generation can be different from the frequency in the parental generation because of chance.

As another simple example, imagine that you have type MN blood for the MN blood group. This means that you have two alleles, an M allele from one parent, and an N allele from your other parent. When you reproduce, you pass on one of these alleles in a sex cell (sperm for males, eggs for females). Is the M allele or the N allele passed on in any given sex cell? It is a 50 : 50 chance, just like flipping a coin to get heads or tails. If you have four children, you might expect to pass on the M allele half the time and the N allele half the time, just as you expect to flip a coin four times and get two heads and two tails. However, while this is the expected outcome for a very large number of cases, by chance you might easily wind up passing the M allele to three children and the N allele to one child, or passing on the M allele to all four children. There are five different outcomes, and statistical theory[3] allows us to predict the probability of each outcome.

Four children inherit your M allele and none inherit your N allele = 1/16 = 0.0625
Three children inherit your M allele and one inherits your N allele = 4/16 = 0.2500
Two children inherit your M allele and two inherit your N allele = 6/16 = 0.3750
One child inherits your M allele and three inherit your N allele = 4/16 = 0.2500
No children inherit your M allele and all inherit your N allele = 1/16 = 0.0625

Now, consider this process happening for everyone in a population. The result is that the allele frequency among the children may not be the same as that of the parents. This is genetic drift.

Computers can be used easily to simulate the process of genetic drift[4]. Figure 1.1a shows an example of genetic drift in a simulated population of 50 individuals (25 couples) for 100 generations. I set the frequency of a hypothetical allele equal to 50 percent in the starting generation. In each generation each couple has two children, replacing themselves so that the population size remains constant at 50 adults each generation. (This restriction is not necessary, but makes the impact of genetic drift easier to see.) The computer program uses randomization to determine what alleles are passed to the next generation, much like flipping coins and counting the number of heads. In the first generation, the frequency changes from 50 percent to 47 percent due to random chance. In the second generation, the frequency remains constant at 47 percent, again due to random chance. In subsequent generations, the frequency drops to 40 percent, then rises to 48 percent, and then rises to 58 percent in the fifth generation. As shown in Figure 1.1a, the frequency of the allele drifts, sometimes going up, sometimes going down, and sometimes staying the same. The direction of change is random.

Because genetic drift is a random process, we could not predict the exact form of the graph in Figure 1.1a, although probability theory allows us to predict that there would be a fair amount of drift (though not the direction) because the population size is small. The random nature of drift also means that each time we conduct a simulation experiment

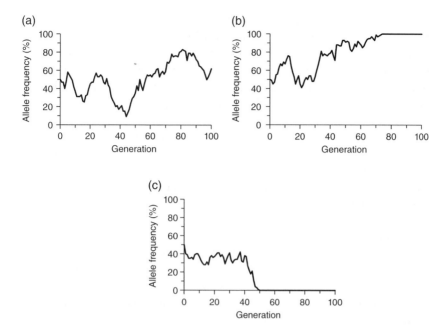

Figure 1.1 Three simulations (a, b, c) of genetic drift over 100 generations in a population of 50 reproductive adults, starting from an initial allele frequency of 50 percent. Genetic drift results in a random fluctuation of allele frequencies over time. Each time the simulation is run, a different picture results—compare Figures 1a, 1b, and 1c.

we are likely to get a different result, just as flipping a set of coins is likely to give us different outcomes from trial to trial. To illustrate this, Figure 1.1b shows a simulation of drift using the same exact starting point—50 adults and an initial starting frequency of 50 percent. In this case, the allele frequency drifts up and down, but in a different pattern, and drifts up to a frequency of 100 percent in the 74th generation. After this point, there is no further change because all the alleles in the population are the same. Figure 1.1c shows yet another example, although in this case the frequency drifts down to zero after 50 generations. Although we cannot predict beforehand the exact path of genetic drift in any specific example, probability theory does allow us to make some basic predictions. First, if enough time goes by, the frequency of an allele will ultimately drift up to 100 percent or down to 0 percent. Second, drift shows the greatest fluctuations from one generation to the next in small populations (just as it is more likely to get seven heads out of 10 flips of a coin than 700 heads out of 1,000 flips of a coin).

Geneticist Motoo Kimura extended the finding of genetic drift to his neutral theory of molecular evolution, which looks at the interaction of mutation and drift. Under this model, most neutral mutations are likely to be lost quickly due to drift—they are so rare to begin with that the odds are against them being passed on to the next generation. However, not all mutant alleles will be lost due to drift. By chance, some mutant alleles will drift up in frequency and become established in a population in the absence of natural selection. The neutral theory does *not* negate evolutionary change due to natural selection, but instead shows that evolutionary change need not always be adaptive. Evolutionary biologists debate the relative impact of drift and selection, but agree that both operate in populations in the real world.

For our purposes here, the take-home lesson is that not all evolution has to be adaptive. Some traits have evolved because of adaptation via natural selection and others are likely to reflect the balance between mutation and drift. In any specific case, such as the traits we will examine for human evolution, we need to examine all available clues to determine if the evolutionary change we see is primarily adaptive or nonadaptive in nature.

Myth #4 In evolution, bigger is always better

Status: This is a misconception. In a popular application of the idea of "survival of the fittest," we tend to equate larger size as having the greater chance of evolutionary success because we assume biggest is the most fit. Although there are indeed many cases where larger individuals have a greater chance at survival and reproduction, there are also cases where being smaller gives one an evolutionary advantage. It all depends on the specific environmental circumstances.

We often view the universe around us in terms that are familiar to us from our daily existence. An example is the tendency to view groups of animals as families even when their social structure is not equivalent to the nuclear family with which we are all familiar. Such misconceptions are particularly common when considering the nature of evolution. One such misconception is the idea that "bigger is better."

There are many examples of people considering that "bigger is better" in different aspects of our life. We can see this principle when buying a computer and choosing the size of the hard disk or active memory. We know from experience that larger disks provide more rooms for all of our files and that more computer memory often allows our programs to run faster. We can see the same principle when shopping at the market. For example, a larger box of breakfast cereal is more cost-effective than a smaller one so that, in terms of our budget, bigger is indeed better. Moreover, we often prefer larger items, be it automobiles, televisions, or diamond rings, for a variety of reasons.

Let us explore the extension of the basic idea of "bigger is better" to the biological world. Many people think of natural selection in terms of the phrase coined by the nineteenth-century sociologist, Herbert Spencer, that natural selection is "survival of the fittest." Is this phrase accurate? The answer depends on the exact use of the word "fittest." Often, the word conjures up an image of traits related to physical fitness, such as size, strength, and speed. Thus, when we say "survival of the fittest," we may picture a situation where the largest, strongest, and fastest individuals are the most likely to survive and reproduce because their physical attributes make them better competitors for mates and food and better able to defend themselves. Given competition for mates or food, it seem reasonable to assume that the largest individuals will be best able to compete, and who in turn would pass on their genes more frequently to the next generation, leading to an evolutionary trend over time where the organisms become larger and larger.

Being bigger need not refer only to overall body size. A somewhat different example comes from examining the fossil record of human evolution, where the average brain size in the genus *Homo* has increased almost 60 percent in the last 2 million years.[5] Because of the association of brain size in different ancestral species with various technological and cultural achievements, we see modern humans as more advanced by virtue of our larger brain, implying that bigger is again better. More will be said on the evolution of brain size in our ancestors in later myths (Myths 22 and 47).

In order to examine the idea that bigger is always better, we need to think in terms of pluses and minuses, costs and benefits. A large car may appeal to us in terms of available space, ruggedness, and speed, but the downside may be lower gas mileage. An 80-inch television might be great for some spaces, but could be overwhelming and difficult to watch in a small room. Large boxes and cans of food can be cheaper per unit cost, but might be more difficult to store. A large house might be desirable, but not affordable. The point here is that we need to consider both the pros and cons of any of the above purchases. In net value, bigger may not always be better.

The same is true of biological phenomena. Larger body size may have an advantage in terms of strength and competition for mates and food, but has the disadvantage of requiring more food energy. Larger body size can thus have both a benefit and a cost. Natural selection operates to lead to a balance between the benefits and costs to maximize fitness. Here, we use the word fitness in the more precise evolutionary context as the probability of survival and reproduction. This probability reflects the net balance between benefits and costs. As an analogy, consider the benefits and costs of advertising in a business. If you own a small business, you have to spend money to make money by investing some funds for advertising. The benefit of advertising is that you will increase the pool of potential customers that otherwise might not be aware of your products or services. The problem is that it costs money to advertise. There is a balance between these benefits and costs. If you spend too little on advertising, you will not reach as many potential customers as you would with a larger advertising budget. However, you would not want to increase the advertising budget without limit; after a certain point, you may saturate your potential market such that more money spent on advertising will not necessarily increase the number of customers. Further, you do not want to keep increasing the amount spent on advertising until the point where it costs more than your profits! It is clear that there is a balance here, and you want to find the sweet spot that maximizes profit and minimizes expense.

Natural selection can be thought of in a similar manner. Body size, for example, can be related to both benefits and costs, which in turn affects overall fitness. If the benefits of larger body size outweigh the disadvantages of larger body size, then natural selection will favor larger size. This will certainly be the case in some environmental contexts, but not all. In cases where available food resources are limited, it might actually be better to be smaller because of lowered energy requirements. The actual fitness of large or small body-sized organisms thus depends on the specific environmental context and shows that bigger is not always better.

What about the example offered earlier concerning the increase in brain size in human evolution? Although the relationship between brain size and cognitive ability and fitness is often complex, we do see a general trend in the fossil record for increasing brain size over the past 2 million years (see Myth 22). At first glance, this trend appears to fit the idea of "bigger is better," which then leads to the common science fiction misconception that future human species will have increasingly larger brains, culminating in the absurd notion that one day humans will resemble giant brains with tiny vestigial arms and legs. A classic example of this notion is found in the entertaining and thought-provoking episode

"The Sixth Finger" from the 1960s science fiction television series *The Outer Limits*.[6] Here, a scientist feeling guilty about his role in the development of nuclear weapons constructs a machine that will allow humans to evolve into an optimistic future state where violence and war have disappeared. Over the course of several treatments, the young man who volunteers to be subjected to artificial evolution changes through a series of increasingly future species. One change is the appearance of a sixth finger, accounting for the title of the episode. The major change, and one expected by virtually any science fiction fan, is the increased size of the brain, ultimately reaching a point where it is obvious to the viewer that the actor had difficulty balancing such a large prosthetic on his head. Accompanying this increase in brain size was an incredibly enhanced intelligence, the development of telepathic and telekinetic powers, and, ultimately, a state of peace and serenity.

Although the specifics of mental abilities in this story are fictional, it is a common idea that our brain will continue to grow in size into the future, a notion tied in with the myth that bigger is necessarily better. In order to consider the relationship between brain size and fitness, we also have to look at the costs of larger brains. For one thing, larger brains consume more energy. The brain is a very expensive organ, requiring 20 percent of our total metabolic energy, a figure much higher than in other mammals.[7] Second, larger brains are harder to cool, because a basic biophysical property of mammals is that larger bodies, limbs, and heads lose heat more slowly than smaller ones. Finally, our species has a limit on how much brains can grow. A certain amount of rapid brain growth in humans takes place before birth, and giving birth to large-brained babies can be hazardous to the mother. If brain growth continued in human evolution, there would be a time when any further advantage of larger brains was offset by the added disadvantage of larger brains. In other words, the costs would exceed the benefits. All other things being equal, natural selection tends toward an optimal balance between benefit and cost. In terms of human brain growth, it is interesting that we already may have peaked in brain size. Biological anthropologist Christopher Ruff and colleagues found that absolute brain size in *Homo sapiens* has actually decreased slightly over the past 35,000 years, in part the result of a similar decline in average body mass.[8]

Perhaps one of the best counterexamples to the myth that bigger is always better is a phenomenon known as island dwarfism, named for the finding that a number of large-bodied species trapped on islands or other isolated areas often show a reduction in body size over time. One of the more spectacular examples of island dwarfism is the fossil remains of dwarf elephants found on islands around the world, and some of these extinct species are estimated to have weighed as little as 200 kilograms (441 pounds). A reasonable evolutionary explanation for island dwarfism is that in some cases of isolation (such as on an island) available food resources are limited and animals that are smaller actually have a better chance of surviving such limitation because smaller bodies require less food. (A possible example of island dwarfism in human evolution is described in Myth 35.) It is also interesting that the opposite pattern, known as island gigantism, sometimes occurs to initially small species, such as some birds and rats, move into an environment lacking predators. The complex factors affecting body size in relationship to food resources, predator–prey relationships, and population growth are beyond the scope of this book, but the main lesson here is that the evolutionarily optimal body size will depend on specific conditions and will not always lead to larger body size.

We see that the idea of "bigger is better" is sometimes true, but it is not an absolute and is very much dependent on the specific local environment to which a species adapts. Sometimes smaller is better. A broader implication of this discussion, which surfaces repeatedly in this book, is the concept that evolution through natural selection represents a compromise between costs and benefits of evolutionary change. We will see a number of examples in human evolution in later myths that reinforce this basic principle that, when it comes to evolution, nothing is free.

Myth #5 Natural selection always works

Status: One common misconception about evolution is that natural selection always works, and a species will always be able to adapt to changing environmental circumstances. This is not the case, and the fact that over 99 percent of all past species are now extinct shows that over the long term natural selection does not continue to work. Because new species are born at the same time that old species die, the process of life continues, but with new players over time.

When I was a teenager in the late 1960s, ecology and environmental issues were on many people's minds. I recall hearing someone claim that we should not worry too much about environmental change because in the long run humans would adapt to a polluted planet. I have always taken this statement as faith in the ingenuity of humans that we will eventually learn to filter out toxins and develop clean energy sources.

Years later, I wondered if this comment actually reflected a belief in the power of biological evolution. I have found that some people have a very optimistic (though undeserved) faith in the ability of natural selection to solve all problems.

Natural selection *is* a remarkable process, but it is not perfect. Selection leads to an optimal solution in terms of the differences in survival and reproduction, but this does not mean it will lead to a perfect solution. As an analogy, consider economic competition between different companies. (I choose this as an analogy because economic competition is often viewed as an analog of the idea of "survival of the fittest.") When one company outdoes another in a fair market competition, it does not have to be perfect, but just better than the competition. Likewise, natural selection will favor those individuals that are better at surviving and/or reproducing than others are, but that does not mean the winners will be perfect. Evolution provides us with countless examples of adaptations that are far from perfect, but instead are good enough and, consequently, involve compromises. An example from human evolution is the tradeoff between the benefits and costs of walking on two legs (covered in more detail in Myth 15). Walking on two legs is not a perfect solution but one that is good enough because it provides a net advantage.

When considering the so-called perfection of natural selection, we also need to acknowledge that there are times when it is not possible to adapt to new environmental conditions. There is no guarantee that natural selection will save a species if conditions change. For one thing, some solutions might be biologically impossible. If, for example, a habitat is flooded, air breathers cannot all of a sudden evolve gills from lungs. Instead, they drown. There are also constraints on all life, such as the need of animals to eat, and a reduction in food resources cannot result in the evolution of animals becoming able to subsist on sunlight.

Even when adaptations are in principle biologically possible, selection has to operate on the variation that is present, and if the variation is not present, then selection will not take place. For example, natural selection has acted on a number of insect species to give them resistance to pesticide. If these populations did not possess the genetic variants allowing resistance, they would be out of luck. New alleles are introduced into a population by mutation and gene flow, but both are independent from the need to have the allele. For example, if you are an insect species and do not have the genetic variation for pesticide resistance to begin with, your need for it cannot make it materialize out of nothingness. Mutation is a random process that is blind to the need for certain mutations to develop when they are needed. Even if the necessary mutation is present, it may often take a long time to increase the frequency of a new allele to a level high enough to result in major changes in survival. If environmental conditions change too fast, a species' ability to adapt through natural selection may be compromised.

Thus, there are many times a population cannot adapt to changes in the environment through the process of natural selection. If the change is severe enough, or occurs too quickly for a species to adapt and recover, it can become extinct. Here, the species has failed to adapt. The extinction of a species is actually very common and happens much more than most people think. An examination of the fossil record shows that of all the species that have ever lived, over 99 percent of them are now extinct.[9] This very large number is a good demonstration that over the long term natural selection may not keep pace in a species as the environment changes.

Extinction happens all the time as species fail to adapt to changing conditions. Incidentally, I do not view the fact that extinction occurs all the time as justification for human practices that increase the rate of extinction (i.e., "They would have died out anyway"). Paleontologists refer to the ongoing process of extinction as background extinction and contrast it with times in earth's history where the extinction rates increased dramatically and were widespread, known as mass extinctions. There have been five mass extinctions in the history of our planet, the most famous of which was the K/T (Cretaceous/Tertiary) extinction that took place a little over 65 million years ago and wiped out a number of plant and animal species, including the dinosaurs. The largest of the five mass extinctions occurred at the end of the Permian period of the Paleozoic Era a bit more than 250 million years ago, when between 80 and 90 percent of species then in existence became extinct.[10] Possible causes include an impact event, severe volcanic activity, and other natural events. In times of catastrophic environmental change, natural selection may not be able to help.

Extinction is not rare and on occasion can wipe out the majority of species. Given the extremely high rate of extinction in the earth's history it might seem a wonder that there are any species still alive! It is also hard to reconcile the high rate of extinction with the fact that there has been an increase in diversity over time. This seemingly paradoxical view is resolved by remembering that as some species die out, others rise to take their place.

Consider an analogy with the number of humans alive on the planet right now, which is a bit more than seven billion. How long will these people live? Let's consider the oldest documented age of a human being, which was a French woman who died in 1997 at 122 years of age. If we take this number as the maximum age of any human, we can safely say that *every* human on the planet today, including any babies born in the time you take to read this sentence, will be dead 122 years from now. Barring any medical miracles, we expect that every person alive right now will be dead before the middle of the twenty-second century. Do you expect that if we were suddenly able to travel into the future to that time that there would be no humans alive? Apart from imaging a scenario of global destruction, I do not think so, for the simple reason that before that time *new* people will be born that replace those who die in the interim. Like all organisms, human beings age and die, but many reproduce before dying, so that life continues.

The same thing applies to the fossil record. Old species die off and new species are born, a process described in more detail in Myth 8. Life undergoes a constant replenishment, and over time we see evolution move in different directions. There is no inevitable direction that natural selection leads to; conditions continue to change, and natural selection does not always work. Species do not live forever. Still, life goes on.

Myth 6

Some species are more evolved than are others

Status: It is common for people to think of some species as being "more evolved" than others, and to further rank species from less evolved to most evolved, with humans typically placed at the extreme position of most evolved. However, most definitions of "most evolved" rely on arbitrary characteristics that reflect our own biases of worth and value. From a purely evolutionary sense, all life shares a common origin and all species, by definition of evolutionary time back to a common ancestor, are equally evolved.

Which animal is more evolved—an ant or a chimpanzee? Given this choice, I imagine that most people would choose the chimpanzee. If the choice was between an ant and a human, I suspect that virtually everyone would argue that humans are more evolved. In fact, the same is likely to be true no matter what organism we compare humans with, be it trees, ants, birds, or chimps. We see similar placement even when we are not talking about evolution. Humans have long been considered the ultimate among living creatures, whether we try to define this position in an evolutionary scheme or not. For example, the primacy of humans in existence is clear from Psalm 8 in the Bible, where humans are described as "a little lower than the angels" and created by God to "have dominion over the works of thy hands." These statements point to the idea that humans have a special place in the universe. This specialness is of course defined here in a spiritual sense, but it is not uncommon to see similar thoughts when considering the biological nature of humanity.

The ranking of living creatures, and the subsequent high status of humans, is an old idea in Western thought. One early example of this type of thinking comes from the Greek philosopher Aristotle, who proposed that living creatures could be arranged in a linear sequence according to various criteria. His system, the *Scala Naturae*, or Ladder of Being, ranked organisms from the simplest to the most complex. At the base of this ladder, Aristotle had lower and higher plants, followed by sponges and jellyfish, and then by other invertebrates, and finally, by fish, amphibians, reptiles, and mammals. Humans were placed at the top of the ladder (after whales).[11] Although some might quibble with the specific placement of some creatures on the ladder, many agree with the basic notion that humans represent the most complex living creature. This view has been incorporated into popular views of evolution, where humans represent the most evolved of all species.

Such schemes seem at first intuitively obvious, but is this because of some special characteristics that humans possess, or does it simply reflect a bias toward elevating our own species above all others? The key problem here is deciding exactly what is meant by "more evolved." What are the criteria for assessing whether one species is more evolved than another? Size? Strength? Visual acuity? Fertility? Camouflage ability? Any list is going to have a subjective element. For example, if we use criteria such as mathematical ability, technological skill, and linguistic prowess, we will certainly rank humans above other creatures. On the other hand, if we use criteria such as being able to have hundreds of offspring, have higher resistance to radiation, and being able to survive for up to a month without food, then the cockroach would rank higher than humans.

We are back to the basic question of what it means to be "more evolved," as the word *more* implies some feature of evolution that allows comparison between two or more species. Is it possible to compare species in terms of which have changed more? To some extent, we do have a measure of evolutionary change, where traits are referred to as "primitive" or "derived." A primitive trait is one that has changed little since the time of a distant ancestor, whereas a derived trait is one that has changed since the time of a common ancestor. Characterizing a trait as primitive or derived is relative to the species being examined.

For example, when we examine tetrapod species (vertebrates with four limbs), we see that five digits (fingers and toes) is very common, seen in numerous amphibian, reptile, bird, and mammal species, including humans. We also know from the fossil record that the first tetrapods had five digits. The widespread occurrence of five digits throughout tetrapods shows us that having five digits is a primitive trait in tetrapods. However, not all tetrapods have five digits; for example, horses have lost four of the digits and have only a single digit. The loss of digits in this context is a derived trait, showing something that has changed in the evolution of horses (and related creatures) since the origin of mammals.

Evolutionary biologists distinguish between primitive and derived traits as a way of determining evolutionary relationships to allow us to reconstruct the evolutionary history of species. However, they are not meant to rank species on a continuum from less to more evolved. For example, if we used absence of five digits in tetrapods (a derived trait) as a measure of how evolved a species is, then we would conclude that horses are more evolved than humans, something that runs counter to what people consider as a ranking of "most evolved."

If we find a species that has a large number of primitive traits, we might call it a primitive species relative to related species. Consider, for example, different kinds of primates, the group of mammals that include monkeys, apes, and humans, as well as tarsiers, lorises, and lemurs. Lemurs are a form of primate that are in a number of ways more similar to the earliest known primates than other forms of primates, such as monkeys and apes. When discussing primate biology and evolution, we might want to summarize the relative difference in evolutionary change by referring to lemurs as more primitive than monkeys or apes. In some features, lemurs have changed less than other primates have. However, does this therefore imply that they are less evolved? No, because by doing so we are implying that evolution can be measured in terms of the amount of change. Evolution is a process that can include different rates of change, including at times a lack of change. For example, in terms of natural selection, a species can evolve in a new direction (say an increase in the size of teeth) to adapt to a new set of environmental conditions, a process known as directional selection. On the other hand, selection might lead to the status quo being maintained, a process known as stabilizing selection, when changes away from the average might reduce fitness. (For example, birth weight in humans, which can be hazardous if the baby is too small or too large.) Both directional and stabilizing selection are different ways in which one of the evolutionary forces (natural selection) can play out, but both are part of the evolutionary process.

The difference is that one leads to directional change and one leads to stability. Using the actual amount of change is therefore not a measure of whether one species is more evolved than another. All species evolve, and the speed at which any given trait changes, while interesting in its own right, is not a relevant measure that can be used to rank species.

What about looking at how long a species has existed? Is it possible to rank species according to their longevity rather than the degree of change so that we can then state that older species are more evolved than younger species? Again, some select examples could provide unflattering contrasts that most people would disagree with, because they do not wind up placing humans as the most evolved. For example, what happens when we compare our species, *Homo sapiens*, which has been around about 200,000 years, with polar bears, which were around 600,000 years ago?[12] Should we conclude that polar bears are more evolved by virtue of having been around three times as long as modern humans? Again, this does not match up with most people's ideas of what "more evolved" means.

We cannot use the longevity of a species as a measure of the amount of evolution because evolution does not start or stop with the divergence of a new species. As will be shown in later myths, *Homo sapiens* arose from an earlier species, *Homo heidelbergensis*, which in turn arose from a still earlier species, *Homo erectus*, and so on into the past. We can do the same for any living species; as we go back in time, we will see different ancestral species. At some point in the past, any two species will have a common ancestor. For example, if we trace both humans and chimpanzees back, we find from genetic and fossil evidence that both lines share a common ape-like ancestor about 6 to 7 million years ago (see Myth 10). At that point, we see that both the human and chimpanzee lines date back to the same point, and the evolutionary lines have equal longevity. We can extend this to all species and, given evidence that all life shares a common origin, we see that no single line has been around longer than another has been.

In truth, there is no way to rank one species above another in terms of how evolved they are relative to each other. The question itself has no direct meaning in terms of how the process of evolution works any more than dropping two identical objects and talking about one being more affected by gravity. When we talk about some species being more evolved than others, we are actually talking about comparison of specific biological and behavioral traits, all of which have an evolutionary history. Thus, I have no difficulty arguing that humans show the greatest achievements in problem solving and technology, but even though these abilities have an evolutionary origin does not mean that we can be characterized as "more evolved." *Differently* evolved would be more accurate.

Myth #7 Humans lived at the same time as the dinosaurs

Status: Some cartoons and movies show modern humans and dinosaurs living at the same time, and some polls have shown a substantial percentage of Americans believe this to be a fact. In reality, dinosaurs first evolved over 200 million years ago, but died out (except for birds) 65 million years ago. Our first bipedal ancestors did not appear in the fossil record until only 6 to 7 million years ago, showing absolutely no overlap in time. Modern geoscience provides many ways of deriving the dates of fossils, which allows us to reject human–dinosaur coexistence and gives us an accurate history of life on earth.

I recall an evening when I was a young child in the early 1960s, when my older brother had a number of friends over at our house clustered around our small black-and-white television set to watch an episode of the animated show, *The Flintstones*, one of the first prime-time cartoon series. *The Flintstones* focused on the lives of two working-class suburban families, but it was set in the Stone Age. One of my favorite parts of the show was how humans and dinosaurs interacted, including the domestication of dinosaurs for labor and as pets. Although this was for me a fun cartoon (and shown in prime time), I never considered it to be a realistic depiction of what life was like for our Stone Age ancestors. Indeed, I recall a brief unit in first grade a few years before where we were shown filmstrips about dinosaurs and it was made very clear to all of us that the dinosaurs lived and died a *long* time before humans were on the scene.

Over the years, I can think of many books and movies that had both humans and dinosaurs coexisting. Sometimes this coexistence was explained by dinosaurs surviving in "*The Lost World*," the title of a novel by Sir Arthur Conan Doyle. Other plots involve time travel (including the classic story, *A Sound of Thunder*, by Ray Bradbury) or cloning of extinct dinosaurs (the book and movie, *Jurassic Park*). Sometimes the coexistence is not explained, as in the 1966 remake of *One Million Years* BC, a date much too young for dinosaurs and too old for the modern humans in the film. These inaccuracies can be ignored in this context because it is fiction, and a large part of fictional enjoyment involves the temporary suspension of disbelief.

However, not everyone seems to realize that the idea of humans and dinosaurs coexisting is fiction and not fact. For example, one survey reports 40 percent of respondents agreeing with the statement "Dinosaurs lived at the same time as people." (An additional 13 percent were not sure.)[13] Further, there have been arguments for evidence of human–dinosaur coexistences, such as the claim that human footprints have been found alongside dinosaur footprints in ancient deposits of limestone in the Paluxy River riverbed in Texas. It turns out that these footprints are not human, but are instead misinterpreted dinosaur tracks, and in some cases have been altered.[14]

It has long been known that dinosaurs lived and died long before the first human-like creatures appeared. Geologic strata containing dinosaur remains occur below those containing remains of early humans. The geologic principle known as the Law of Superposition states that layers of sedimentary rocks are deposited over time such that the older strata are below younger strata. Dinosaurs are found in the geologic time period known as the Mesozoic Era, which is subdivided into three geologic periods: the Triassic, Jurassic, and Cretaceous periods. Dinosaurs appeared during the Triassic period, became dominant during the Jurassic period, and died out at the end of the Cretaceous period. The Mesozoic Era was followed by the Cenozoic Era, often referred to as the Age of Mammals because it is in this time that modern groups of mammals evolved from earlier mammals. The Cenozoic Era is divided into three periods, which are divided into seven geologic epochs. The first mammals included some ancestors of later primates, but no humans (or monkeys or apes). Over time, primates diversified, and human ancestors, the first bipedal apes, do not appear until much later, toward the end of the fourth of the seven geologic epochs (the Miocene epoch) of the Cenozoic Era.

Although we have long known that dinosaur fossils are older than human fossils, we did not always know exactly how much older. Using modern geologic methods that have been developed since the middle of the twentieth century, we can now provide actual dates to events in earth's history. The start of the Triassic period is now dated to 252 million years ago, the start of the Jurassic period to 201 million years ago, and the start of the Cretaceous period to 145 million years ago. The end of the Cretaceous period, when the dinosaurs and many other species died out, is dated to 66 million years ago.[15] Primates, including humans, evolved after this time. The earliest evidence of the first bipeds (human ancestors, but not modern humans) dates to at least 6 million years ago and perhaps as much as 7 million years ago (see Myth 10). The first fossils assigned to the same genus as us (*Homo*) date to a bit more than 2 million years ago.

The first modern humans date to 200,000 years ago. It is clear that dinosaurs and humans, even the earliest human ancestors, do not overlap in time with dinosaurs, or come anywhere near it.

A reasonable question here would be—how do we know these dates? Although the Law of Superposition gives us relative ideas of age (which fossils are older), there are a number of methods that provide us with precise estimates of age.[16] The oldest of these methods to be discovered was carbon-14 dating, developed by chemist Willard Libby in 1949, an accomplishment for which he was awarded a Nobel Prize. Carbon-14 dating is one of several dating methods that use the principle of radioactive decay. During their lives, all living organisms (including us), absorb ordinary carbon (^{12}C) as well as the radioactive isotope carbon-14 (^{14}C), which has two more neutrons. Both forms of carbon are found in the atmosphere, but when an organism dies it no longer absorbs carbon, and its carbon-14 begins to decay into nitrogen-14. This decay occurs at a geometrically decreasing rate, known as a half-life, which is the amount of time it takes for half of the radioactive carbon-14 isotope to decay. The half-life for carbon-14 is 5,730 years, which means that after 5,730 years, half of the carbon-14 is gone, and it takes another half-life for half of the remaining carbon-14 to decay. Thus, it takes *two* half-lives ($5,730 \times 2 = 11,460$ years) to have 75 percent of the carbon-14 to decay. (One half-life = 50 percent and the second half-life is half the remainder, 25 percent, adding up to 75 percent.) After three half-lives ($5,730 \times 3 = 17,190$ years), half of the remaining 25 percent is gone, and $75 + 25/2 = 87.5$ percent of the original carbon-14 has decayed. In practice, geologists measure the amount of radioactive emissions in a sample and compare it to the rate in living organisms to determine how many half-lives have gone by, thus giving us a date.

Like all dating methods, carbon-14 dating has several restrictions. First, it can only be applied to organic matter, such as bone, wood, or charcoal from a fire. It would be useless for dating rock. Second, the amount of carbon-14 in the atmosphere had fluctuated in the past due to climate change, although calibration curves are available to adjust for this problem. Third, the half-life for carbon-14 is short, which means that the amount of carbon-14 in a sample will be very small after a few half-lives, until it is too low to measure accurately. In general, carbon-14 dating is good for samples from the last 50,000 years. We do not use carbon-14 to date the last days of the dinosaurs. However, many other dating methods have been discovered followed the invention of carbon-14 dating, and these methods extend our ability to measure geologic dates.

One widely used method is argon dating, which also involves the principle of radioactive decay. As originally developed, argon dating makes use of the decay of an isotope of potassium (^{40}K) into argon gas (^{40}Ar). Because the half-life is very long (1.25 billion years), this method is very useful for dating events older than about 100,000 years. A variant of the method is often used that first converts a different isotope of potassium (^{39}K) to an argon isotope (^{39}Ar), which turns out to be more accurate. Argon dating allows geologists to date volcanic rock because argon gas builds in volcanic rock over time. When a volcanic eruption occurs, the molten lava is very hot and any existing argon gas is driven out, thus resetting the clock. After the lava has cooled, the process of radioactive decay begins again, allowing us to measure how much radioactive argon accumulates over time, which provides us with the date of the volcanic eruption. This method is useful for dating fossils of organisms that died between two volcanic eruptions. For example, if we find a fossil above one layer of volcanic rock dated at 16 million years and below a layer dated to 16.2 million years, we know that the fossil died between those dates. Argon dating has been very useful in studies of early human evolution in East Africa, a region that was very volcanically active in the past.

Obviously, argon dating is not useful in cases where there have been no volcanic eruptions, but there are other dating methods that could be used in different cases. For example, uranium-lead dating also uses the principle of radioactive decay, has a very high half-life, and can be applied to a number of different minerals. Still other dating methods make use of different physical properties besides radioactive decay. Fission-track dating is a method that counts the number of tracks left in volcanic glass when uranium decays. Electron spin resonance dating is a method that utilizes the number of radioactive atoms in calcite crystals found in bones and shells. Another method known as thermoluminescence dating can provide dates based on the accumulation of electrons in objects that have been heated, including, for studies of human evolution, pottery and other objects.

There are also dating methods that make use of correlations established from other dating methods. One example is biostratigraphy, which compares distributions of animal bones from different sites to get an idea of the age of a given site. Suppose, for example, you see a particular anatomical trait in a species that you know from other dating methods only lived 3 million years ago. If you find this trait in fossils at a site where other dating methods are not available, it is a pretty good inference that this particular site is also dated around 3 million years ago.

One of the more elaborate correlational methods makes use of the fact that the magnetic pole of the earth has changed from being at the north magnetic pole to the south magnetic pole periodically in earth's history. For example, a compass would point to the north magnetic pole for the last 780,000 years, but would point south for tens of thousands of years before that. These shifts can be detected in minerals and occur at irregular intervals in the past. Other dating methods have been used to develop a timeline for these reversals, allowing geologists to date a site relative to the calibrated record of past magnetic reversals.

These are but a few of the examples that we can use to date past events in earth's history, including many in the course of human evolution. When possible, multiple methods are used to provide as accurate a record as possible. In the case of the dinosaurs, modern geologic methods have confirmed what had been known all along—the dinosaurs died out a long time before the earliest human ancestor walked the planet.

As a final note, I have to mention that although the classic dinosaurs that we are all familiar with are long gone, a related form of life is still with us—the birds. A number of paleontologists have proposed that birds are descended from avian dinosaurs, which means that some dinosaur descendants are still with us. However, seeing a sparrow on a tree is a far cry from the usual idea of human–dinosaur existence, as we will not be seeing (thankfully) *Tyrannosaurus* or *Velociraptor* walking down the street except in the movies.

Notes

1 See Cook *et al.* (1999) for more details regarding the evolution of the peppered moth.
2 "*Forrest Gump* Quotes." Available at http://www.imdb.com/title/tt0109830/quotes (accessed July 1, 2014).
3 See any college statistics book.
4 See Relethford (2012: ch. 5).
5 This figure is based on a brain size of 850 cubic centimeters (cc) in early *Homo erectus* (Rightmire 1990) compared with an average brain size of 1350 cc in modern humans.
6 Schow and Frentzen (1986).
7 Armstrong (1983).
8 Ruff *et al.* (1997).
9 Raup (1991).
10 Benton (2003).
11 Kennedy (1976).

12 Hailer *et al.* (2012).
13 Bishop *et al.* (2010).
14 Milne and Schafersman (1983).
15 Dates are taken from the version of January 2015 of the International Chronostratigraphic Chart published by the International Commission on Stratigraphy. Available at http://www.stratigraphy.org/ICSchart/Chronostrat Chart2015-01.pdf (accessed June 9, 2015). These dates are periodically revised slightly as more accurate estimates of geologic boundaries are developed. The most recent version at any point in time is available at http://www.stratigraphy.org/index.php/ics-chart-timescale (accessed July 5, 2016).
16 Dating methods are described in a large number of introductory texts in archaeology, biological anthropology, paleontology, and geology.

2 HUMAN ORIGINS

The idea that humans evolved from an early ape-like ancestor is not new to most people. Unfortunately, the simple summary of "apes evolving into humans" is an oversimplification. As will be described in this section of the book, the transition from an ape-like ancestor to human was not instantaneous or simultaneous. There is a long period of human evolution (4 million years or more) where some aspects of the human condition had evolved, but not others. This section of the book examines some of the myths and misconceptions associated with the evolution of the apes and the emergence of early human ancestors.

Myth 8 If apes evolved into humans, then apes should not exist today

Status: This is indeed a myth, based on a misunderstanding of how evolution works. An earlier ape-like species did not "turn into" the human line, but instead the human line split off as a new evolutionary line.

One argument that I have heard used to reject the idea of human evolution goes as follows: "If apes evolved into humans, then there should be no apes in the world today. However, there are apes living today, and therefore humans could not have evolved from apes." The first time I heard this argument I dismissed it as nonsense. After hearing it several times, I realized it was a good question, but one whose answer rested upon incorrect assumptions about how evolution works. Because this argument is so ubiquitous (I have heard it from a wide range of sources),

50 Great Myths of Human Evolution: Understanding Misconceptions about Our Origins, First Edition. John H. Relethford.
© 2017 John Wiley & Sons, Inc. Published 2017 by John Wiley & Sons, Inc.

it is worthwhile examining the underlying assumptions and the chain of logic in some detail.

Let us start with the statement "If apes evolved into humans." Yes, anthropologists state that the human line did evolve from an ape-like ancestor. However, when we use words such as "ape" and "ape-like," we are talking about earlier, primitive ancestors and *not* modern apes. The ape-like ancestor we are talking about is not the same as a modern chimpanzee (see Myth 14 for more details). Let us ignore this point for the moment and just concentrate on the evolutionary implications that some sort of ape evolved into an early form of human, even though this is an overly simplistic statement, because I want to concentrate here on the underlying logic and how evolution works.

The key piece of logic here is the assumption that if a species evolves over time, it has by definition changed and the earlier form can no longer exist. Consider the statement that species A evolved over time into a present-day species B. To many, this means that species A has transformed into species B. Species A lived in the past and species B lives today. Under this model of evolutionary change, species A is no longer around. The final logical piece is to note that if species A is indeed still around, then the evolution of A into B did not occur. In terms of human evolution, the logical conclusion is that because there are still species of apes alive in the world today, we therefore did not evolve from any ape or ape-like species.

The model of evolutionary change implicit in the above argument is one where a species evolves over time into a new species. This is known as anagenesis (see Figure 2.1a). Imagine that species A existed at some point in the past. Over time, evolutionary change will occur in this species due to the effect of natural selection and genetic drift on newly arising mutations. As time goes on, the amount of evolutionary change can accumulate. If we were observing this process in close detail in the fossil record, we might see changes in the anatomy of this species, such as changes in body size, tooth size, number of digits, among many other possible examples. Given enough time, the overall appearance of organisms in this evolving lineage might have changed enough that we would give it a new label, say species B. Here, the change from species A to species B is a gradual process (in geological time). Even though we might use two species names, A and B, we are talking about labels to indicate different stages in the evolution of a single line. An analogy would be our life cycle and how we label different stages (such as infancy, childhood, adolescence, and adulthood), but are always talking about the same person. In this context, we can consider the infant changing into a child over time, and the adolescent changing into an adult over time. We all start life

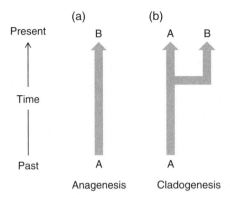

Figure 2.1 Models of species change. (a) Anagenesis, the change over time in a single evolutionary line where species names are given as labels with different points along the line. Here, species A changes over time enough so that at some point paleontologists might label the later version as species B. (b) Cladogenesis, the birth of new species through reproductive isolation and genetic divergence over time. Here, some part of the parent species A splits off and forms a new, reproductively isolated daughter species, B. Note that the two lines (parent and daughter) can continue to live alongside each other.

as infants, but change over time to become children, teenagers, and then adults. We are never both infants and adults at the same time.

Given that anagenesis is the same type of process—change within an evolving line over time—we might expect to see this change play out across a variety of species. We might envision evolutionary history as a series of parallel changes across a number of species. Here, ancient horses evolve into modern horses, ancient apes evolve into modern apes, and ancient humans evolve into modern humans. How then do we go from ape to human? How can both exist today? The problem with this myth is that although anagenesis is part of the evolutionary process, it is not the only part. Evolution does not consist only of an earlier form of a lineage evolving into a modern form. Indeed, how could anagenesis alone provide a complete description of the evolutionary process? It could not, because if we postulate that all evolution occurs in this straight-line fashion, we are left with no explanation of where new species come from.

The answer to this dilemma is that evolution often involves the "birth" of a new species, where a daughter species branches off the line of the parent species and then coexists with the old species. From one species, we now have two. As an example, consider a part of species A branching off and becoming genetically distinct over time, thus forming a new species, B. It is then possible for both species A (the parental species) and species B (the daughter species) to live side by side over time (see Figure 2.1b). The common use of terms such as "parent species" and "daughter species" illustrates why this process (known as cladogenesis) can be described as the birth of a new species. You can extend the birth analogy to your own existence. Your mother gave birth to you, and unless she died in childbirth, there is a period of time when both parent and child exist simultaneously. Even though your mother is an immediate biological ancestor that gave rise to you as an individual, she does not cease to exist the moment you are born. This is because she gave birth to you (as in cladogenesis) but did not change into you (as in anagenesis).

One important point remains to be explained, which is how new species appear. What exactly do we mean by the birth of a species? To see how speciation works, we need to start with a brief discussion of the various definitions of the term "species." Sometimes we use species names as labels to denote different stages within the ongoing evolution of a single species under anagenesis. Under cladogenesis, however, the term "species" most often refers to a specific meaning that concerns interbreeding. Under what is known as the biological species concept, two populations belong to the same species if they naturally interbreed and can produce fertile offspring. If these two criteria are not both met, then the two populations are classified as different species, and are considered reproductively and evolutionarily separate. In other words, any evolutionary change that occurs in one species will not affect other species because they are by definition incapable of exchanging genetic material. Sometimes these differences are immediately clear; there is no need to think too hard about gorillas and goldfish as being separate species because the mere thought of them being capable of interbreeding is ludicrous. On the other hand, consider the classic example of the horse and donkey. These populations are very similar equines and are capable of interbreeding. They can in fact produce offspring (known as mules), but these offspring are themselves mostly sterile and incapable of reproducing with each other. Under the biological species concept, the horse and donkey are different species because they do not produce fertile offspring.

The formation of a new species is a function of the interplay of the four evolutionary forces described in an earlier myth (Myth 3): mutation, drift, selection, and gene flow. Key here is the role of gene flow in connecting populations to each other. Remember that gene flow is the glue that holds a species together; given enough gene flow, genetic changes that occur in one population are shared with other populations. What happens in one population is likely to affect another population when enough gene flow is present. When gene flow is eliminated or reduced greatly, populations are genetically isolated and are free to change independently. A mutation may pop up in one population, but not another. Selection and drift can then affect the frequency of the mutation in different ways.

Populations can be genetically isolated in a number of ways, including the very common phenomenon of geographic isolation—if two populations wind up far from each other, they are less likely to share genes. Over time, the processes of mutation, selection, and drift will act to make the isolated population genetically different. If this isolation occurs for only a short time, the impact may be negligible. However, over time, the daughter population is likely to become more and more genetically different from the parent species. If enough time goes by and sufficient genetic change occurs, then interbreeding may not be possible, or be less fruitful.

The description here of speciation is necessarily simplistic (entire books have been written about it) and the process is not always as clear as might sound. For example, there are examples where populations are geographically and reproductively isolated for the most part and are generally recognized as different species, but are still capable of interbreeding. (A good example from primate biology is the hybridization of different baboon species.) Given enough time, however, the genetic differences can accumulate to reach the level of difference between horse and donkey. Given even more time, the differences might be even greater, such as between ape and human, and given even more time, the differences may be of the order of gorillas and goldfish. In general, the longer two evolutionary lines have been separate, the greater the difference in both genetics and anatomy.

At some point in primate evolution, the evolutionary line eventually leading to humans became evolutionarily separate through this process of cladogenesis. Before 6 to 7 million years ago, there were many different species of primates that we often refer to collectively as apes but none that we would consider human or human-like. Since that time, cladogenesis (and extinction) has occurred, leading to ever-changing patterns of diversity. Species died out and new species arose. Today, we have apes and humans in our world. Because evolution is not just anagenesis, the presence of apes today does not negate that fact that our ancestry lies with earlier populations of apes.

Studies of the fossil record of many organisms show that both anagenesis and cladogenesis have occurred in the past and a good case can be made for examples of both in the human fossil record. The main point in this myth is that when anthropologists make the claim that an ape-like ancestor evolved into a human-like ancestor, they are *not* claiming that a single species of apes (such as chimpanzees) evolved into the human line via anagenesis. Instead, we are stating that modern chimpanzees and modern humans have a common ancestor earlier in time. Nor do we suggest that this common ancestor was itself a chimpanzee. Modern chimpanzees (among other apes) and modern humans are the product of cladogenesis, the birth of new species over time, and our collective family tree thus more closely resembles a bush with many branches rather than a tree with one trunk and no branches.

We will see many examples of new species appearing in human evolution. Even though we have only one human species at present, in the past there were times when there were multiple evolutionary lines. Not all survived to the present. As noted earlier (Myth 5), many evolutionary lines become extinct. Just as cladogenesis is described efficiently as the birth of a new species, so too can extinction be considered the death of a species. As with the history of individuals within a species, evolutionary history can be thought of as a series of births and deaths, and the diversity of life at any point in time is a reflection of this constant turnover in species.

Myth #9 "*Ramapithecus*" was a human ancestor

Status: This is not so much a myth as it is an earlier hypothesis that was popular in the 1960s. This hypothesis fit the data at one time, but additional evidence rejects it, and shows that "Ramapithecus" *was actually a fossil ape that is related to orangutans.*

As noted in the introduction, some ideas that we label as myths relative to our present state of knowledge were once considered accurate depictions before evidence accumulated leading to their rejection. This myth deals with one such example—the interpretation of fossils known as "*Ramapithecus.*" (The reason this name is in quotes is discussed later in this myth.) "*Ramapithecus*" figures in our changing interpretations of the evolutionary relationship between the living apes and ourselves as new data were discovered. As such, the case of "*Ramapithecus*" is an excellent example of the dynamic nature of science described in the introduction to this book. At one time, the evidence supported the view that "*Ramapithecus*" was a bipedal human ancestor. The accumulation of new data has since shown this hypothesis is incorrect, and "*Ramapithecus*" is actually a fossil ape related to modern orangutans.

In order to understand the story of "*Ramapithecus*" and other essays on human origins, we need to review some basic information on the living apes. It is common knowledge, based on comparative studies of genetics and anatomy, that the African apes are our closest living relatives. The living African apes consist of the gorilla, the chimpanzee, and the bonobo. All are similar in a number of physical characteristics, such as having longer arms than legs. The African apes also all share knuckle walking, which is a form of four-legged movement where the front limbs rest upon the knuckles and not the palms. Because of the long arms, the spine of a knuckle-walking ape is at an angle to the ground, unlike the parallel orientation of the spine of animals with front and rear limbs of equal size, such as a monkey. The three types of African apes also show a number of differences from each other in body size and shape. Gorillas are large and impressive, particularly the mountain gorilla, where adult females averaging about 98 kilograms (216 pounds), and adult males averaging about 159 kilograms (351 pounds). Chimpanzees and bonobos are much smaller.[1]

At some point in the past, we shared a common ancestor with the African apes. Our best current genetic estimates show that we are a bit more closely related to the chimpanzee and bonobo and a little more distant from the gorilla. These genetic relationships (described more in the next myth) support the "family tree" of relationships shown in Figure 2.2. Here, the earliest common ancestor (labeled "A" in the figure) gives rise to two lines, one leading to the living gorillas and the other to the common ancestor of chimpanzees, bonobos, and humans (labeled "B"). At that point, one line leads to the common ancestor of the chimpanzees and bonobos (labeled "C") and the other line leads to living humans. It is this last line that is of particular interest to anthropologists because this line represents the evolution of the hominins, the name we give to ourselves and all of our ancestors back to the time of the divergence from an African ape line. (Some older texts use the term "hominid.") Incidentally, do not be misled by the simple family tree in Figure 2.2 and conclude that hominin evolution consisted of a single straight line—it did not. Instead, there have been a number of side branches to the tree and many times when more than one hominin species existed at the same time (as will be discussed in Myth 18). The simplified view in Figure 2.2 is meant only to show the evolutionary relationships of humans and the African apes.

One of the key questions in anthropology is the identity of the very first hominin. How can we tell if a fossil is a hominin, and thus part of

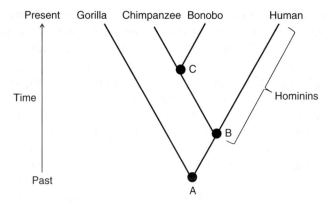

Figure 2.2 A family tree showing the evolutionary relationships between humans and the African apes (gorilla, chimpanzee, bonobo). At some point in the past, a common ancestor (A) gave rise to two lines, one leading to the gorilla and the other leading to the common ancestor of humans, chimpanzees, and bonobos (B). The chimpanzee-bonobo line later split from a common ancestor (C) leading to two separate species. We use the term "hominin" to refer to humans and all ancestors going back to the time of the split from the ape line. As will be shown in a number of myths, the earliest hominins were like us in some ways, but still ape-like in many others. Thus, all humans are hominins, but not all hominins are humans.

our ancestry, or is instead a member of an evolutionary lineage of apes? This is not as simple a question as it might seem at first. It is easy enough to tell the difference between a *modern* human and a *modern* ape. Humans walk on two legs, have larger brains, have small canine teeth, and are relatively hairless. On the other hand, apes walk on all fours, have smaller brains, have larger canine teeth, and are hairy. These differences are quite apparent to anyone who has seen apes at a zoo. However, remember that the differences we see today are the result of millions of years of separate evolution. The further back in time we go in the fossil record, the more ape-like our own line appears. Ideally, we look for evidence of upright walking, which evolved much earlier than large brains (Myth 12), but this is not always available. Sometimes, we look for clues of hominin ancestry in the teeth and jaws. Dental remains are more often found than the rest of the skeleton because jaws and teeth are hard and tough and are thus more likely to fossilize.

As mammals, both apes and humans have four different types of teeth: incisors, canines, premolars, and molars. Incisors are the flat teeth you have in the front of your mouth, four in your upper jaw, and four in your lower jaw. These teeth are used frequently for gnawing, such as when you eat corn on the cob or take a bite of an apple. Behind the incisors are the canines, two in each jaw. In most mammals (including apes), the canines are sharp and pointed, but in humans they are small and blunt. Incisors and canines are the front teeth. The back teeth are the premolars and molars and are used for chewing and grinding food. Structurally, premolar teeth and molar teeth are different, and molar teeth are generally larger. Comparatively, humans and ape teeth are actually quite similar. We have the same number of each tooth (typically two incisors, one canine, two premolars, and three molars in each half of each jaw for adults, for an average of 32 teeth). We also share the same pattern of raised areas (cusps) on our molar teeth, different from the more distantly related monkeys.

However, there are also clear differences between the jaws and teeth of *modern* apes and humans.[2] For one thing, the shape of the jaws is different. In apes, the back teeth are aligned in a straight line, such that the sides of the jaws are parallel to one another, referred to as U-shaped because of the resemblance to the parallel sides of the letter U. For humans, the jaws are curved from front to back, forming the shape of a parabola. A major difference between ape and human teeth is that apes have larger canine teeth that project beyond the height of the other teeth. Ape canines are also sharper and more pointed. Human canines are smaller and blunter and do not project beyond the height of the other teeth. The canine teeth of apes and humans are also used in different ways. Ape canines can be used for slashing and make formidable weapons. Human canines are not as impressive. In fact, our canine teeth are actually very puny and function more like incisors than as nonhuman canines. The reduction in the size of our canine teeth is one of the unique features of human evolution that anthropologists are interested in explaining.

A number of other ape–human differences relate to the size and function of the canine teeth. Because ape canines are large and projecting, they need a space in the opposite jaw for the canines to fit into, otherwise an ape would never be able to close its mouth. When you look at an ape's jaws, you see these notable gaps. In the upper jaw, apes have a gap between the incisor teeth and the upper canine into which the lower canine fits, and there is a gap between the canine and the premolars in the lower jaw into which the upper canine fits. Thus, the two jaws are able to mesh, allowing the ape to close their mouth. The large canine of an ape is also related to a difference in the lower premolar right behind the canine. (Technically this tooth is known as the lower third premolar because the first and second premolars of early mammals have been lost during primate evolution, even though it is the first premolar in the jaw from front to back.) Premolar teeth in humans are also known as bicuspids because there are two raised areas known as cusps. In apes, the lower premolar behind the canine tends to have a single dominant cusp. When an ape closes its jaw, the back edge of the upper canine wears against this cusp, forming what is known as the canine/premolar honing complex, where the back of the canine is sharpened. Humans do not have the honing complex, instead having bicuspid premolars and small canines that wear at their tips, rather than on the back edge.

The dental differences described here came into play in the case of fossils assigned to the genus *Ramapithecus*. In the 1960s, a candidate for the earliest hominin was *Ramapithecus*, the name given to some fossils first discovered in India back in 1932. The genus name *Ramapithecus* translates as "Rama's ape," named after the Hindu deity Rama. The original fossils consisted of two pieces of a broken upper jaw. Though fragmentary, these remains included a number of characteristics that later led some anthropologists to consider that *Ramapithecus* belonged to the human line. The canine teeth were relatively small, the molars appeared similar to humans, and the reconstructed shape of the jaw was parabolic, thus resembling humans more than apes. Furthermore, it was suggested that *Ramapithecus* had a relatively short face, a trait that also distinguishes between modern humans and apes. Additional dental remains were found, including some from Africa that some felt belonged with *Ramapithecus*. By the 1960s, a widely accepted conclusion was that *Ramapithecus* was an early human-like ancestor. The date for the fossils was estimated to be 14 million years ago. This early date implied that the hominin line had split off from the line leading to African apes somewhere before this time.

The focus on the small canines of humans and human ancestors not only helped place *Ramapithecus* as a human ancestor according to some, but also raised the question of the evolutionary origin of small canines. Ever since Darwin, a traditional explanation focused on the replacement of large canines by tools. According to this view, when humans began using stone tools and (presumably) weapons, they no longer needed large canine teeth for defense or offense. Consequently, there was an evolutionary shift from larger to smaller canines. This idea of disuse fell out of favor over time, but there was still a consensus that there was an evolutionary advantage to having smaller canines *if* you used tools. If true, then this link between small canines and tool use would predict that if we found a human ancestor (based on small canines and related dental features), then this ancestor most likely was a tool user. Many anthropologists also considered a link between tool use and bipedalism. This argument was based on the idea that making and using tools is best accomplished by having your hands free, which means walking on two legs. Furthermore, making and using tools suggests an increase in intelligence with perhaps some changes in brain size and structure over the course of evolutionary history.

This entire series of suggested links follows a once-common view that the unique aspects of human evolution all appeared more or less simultaneously. The expectation is that we would see a gradual change from a four-footed, small-brained ape with large canines and no apparent tool use to the emergence of a bipedal, large-brained, tool user with small canines. The use and manufacture of stone tools was seen as the driving force, as tool use was viewed to lead to bipedalism, large brains, and smaller canines. As we will see later (Myth 12), the fossil record has now grown to a point where we see that these changes did not take place all at the same time. However, only a few decades ago we did not have as much evidence, and the tool-use model of human origins could still be defended. In this context, it seemed reasonable then that if we found one piece of the whole picture, such as small canines, we could declare the specimen a human ancestor and fill in the rest of the picture.

Based on the evidence then available, the inference that *Ramapithecus* was an early human ancestor made sense because the dental evidence pointed to greater similarity with humans rather than apes. However, many felt that the most convincing evidence for *Ramapithecus* being a human ancestor was missing—skeletal evidence showing upright walking. Still, the dental evidence convinced many, at least tentatively. Even proponents of the view that *Ramapithecus* was a human ancestor noted that this view was a hypothesis that remained to be more fully tested with additional data.[3]

In later years, the hypothesis that *Ramapithecus* was a human ancestor hypothesis was rejected after the accumulation of fossil and genetic evidence. For one thing, not everyone accepted the conclusion that *Ramapithecus* was a human ancestor, pointing out that the evidence was fragmentary and little was known of the range of variation in characteristics such as canine size to use them for unambiguous classification. For example, female apes typically have smaller canines than males, and sex differences and species differences can be confused. In addition, the conclusion of a parabolic jaw shape was questioned because it was based on inferences made from a broken and incomplete jaw, and some argued that a more ape-like reconstruction (U-shaped) could be made. For some anthropologists, the pieces of evidence that *Ramapithecus* was a human ancestor were falling away, one by one. In addition, the development of methods for using genetic information from living species to reconstruct evolutionary timelines had produced an estimate of the split of ape and human lines to only 4 to 5 million years ago. If true, then *Ramapithecus* could not have been on the human line at 14 million years ago because the human line had not yet split. (See the next myth for more detail on the inferences from molecular anthropology.)

As is often the case, such arguments were settled by additional fossil evidence. As additional Asian fossils were uncovered, it became clear that *Ramapithecus* could actually fit comfortably within the range of variation seen in a previously discovered fossil ape, *Sivapithecus* (named after the Hindu deity Shiva). For example, it turns out that the size of the canine teeth varies even among apes, with some species having smaller canines and, on average, females having smaller canines than males. Small canines by themselves are not always diagnostic of human affinity, and attention has to be paid to other features, such as the pattern of wear on the canines. As specimens of *Ramapithecus* were found to fit comfortably within the range of variation of *Sivapithecus*, it became clear that the two groups were the same, and what had been called *Ramapithecus* was placed into the genus *Sivapithecus*. We now put quotation marks around the genus name "*Ramapithecus*" to indicate that we no longer consider the genus name valid and are using it only for historic context. By the way, the reason "*Ramapithecus*" was subsumed under *Sivapithecus* and not the other way around is that the international rules of classification that all scientists use states that in such cases the earlier published name has precedence.

If "*Ramapithecus*" (= *Sivapithecus*) was not a human ancestor, then what was it? The discovery of parts of a *Sivapithecus* skull from Pakistan in the early 1980s provided the answer to the question of its evolutionary relationship. This skull shows a number of interesting features, such as close-set oval eye orbits and a sloping facial profile. These traits (and others) are important because they do not look like African apes, but instead bear a striking resemblance to the Asian great ape, the orangutan. Additional fossils have shown that *Sivapithecus* lies close to the evolutionary line leading to orangutans and not the African apes or humans. If *Sivapithecus* was an Asian ape, it not only was *not* a human ancestor, but also was not a common ancestor of African apes and humans, because it was already on the line leading to modern orangutans.[4] As such, it is not on any of the lines in the tree shown in Figure 2.2, but on a line that had diverged earlier in time.

You should keep in mind that although the accumulation of fossil and genetic evidence rejected the hypothesis that *"Ramapithecus"* was a human ancestor, that initial hypothesis made sense given the data available at the time. Additional data led to its rejection. As noted in the introduction to this book, this is the way that science is supposed to work. Rejecting incorrect hypotheses helps us converge on the correct explanation.

Humans and African apes split from each other over 15 million years ago

Status: This is not so much a myth as it is an earlier hypothesis that was once favored by the available evidence. Additional genetic and fossil evidence have moved away from an early split of apes and humans and favor a more recent split around 6 to 7 million years ago. Although some have suggested that this date may be a bit older, it would not be as old as 15 million years.

The previous myth noted that our closest living relatives are the African apes, with somewhat greater similarity with chimpanzees and bonobos than with gorillas. In terms of human origins, we once thought that our evolutionary line split from the African apes over 15 to 20 million years ago. The steady accumulation of more fossil evidence as well as developments in comparative genetics show that our common ancestry with African apes is more recent, with most estimates for the evolutionary split being about 6 to 7 million years or so. This shift in dates reflects our growing knowledge of close kinship with the African apes and shows us another example of how fossil evidence leads to the generation of hypotheses that are then tested with additional fossil (and genetic) evidence. The 15 to 20 million-year-old hypothesis for the split of the evolutionary lines of African apes and humans has now been rejected.

The idea of an early (15 to 20 million years old) divergence of the African ape and human lines was supported initially by a number of early discoveries of fossil apes. An important find was an extinct ape known as *Proconsul*, found by the famous anthropologist Louis Leakey and colleagues in Africa, with some specimens dating back to about 20 million years ago. The genus name *Proconsul* translates literally as "before Consul," which makes sense given that Consul was a common name given to performing chimpanzees at the time. The description of this fossil as being ancestral to living African apes was apt, because initial analysis of jaws and apes showed many similarities. *Proconsul* had the typical U-shaped jaw, large canines, gaps near the canines, and single-cusped lower premolar teeth associated with modern apes. In labs, my students have no trouble seeing the ape characteristics of the jaws and teeth of *Proconsul*.

Several different species of *Proconsul* have been identified, and one of the most noticeable differences is size. Some specimens are larger than are others, much in the same way that gorillas are larger than chimpanzees among living African apes. Indeed, this variation in size in living African apes provided context for interpreting the dental remains of *Proconsul*. Today we have bigger African apes (gorillas) and smaller African apes (chimpanzees), and in the distant past, we had bigger and smaller versions of *Proconsul*. Perhaps there was a connection. A model developed where the small species of *Proconsul* was taken to be the direct ancestor of living chimpanzees and the large species of *Proconsul* was the direct ancestor of living gorillas. If we accept this model, this means that the two lines had already diverged almost 20 million years ago.

Another piece of the puzzle came from the discovery of "*Ramapithecus*," described in the previous myth. As recounted there, "*Ramapithecus*" was first considered as a human ancestor and lived 14 million years ago. This meant that by 14 million years ago, the human ancestral line had already split off from the African apes. Taken together with the evidence from *Proconsul* suggesting an early separation of chimpanzee and gorilla lines, the "*Ramapithecus*" fossils suggested that the African ape and human lines had been separate at least 15 to 20 million years ago.

Although the early divergence hypothesis was supported by the data then available, it is important to remember that the data were limited primarily to dental remains. As shown in the previous myth, additional fossil evidence was discovered that removed "*Ramapithecus*" from the human line and relegated it to an ancient relative of the orangutan. Further discoveries and analysis of *Proconsul* also provided a different evolutionary position than that of direct ancestors of living African apes. Although the jaws and teeth of *Proconsul* definitely resembled those of modern African apes, the rest of the skeleton told a different story, showing a mixture of monkey-like and ape-like traits.

Living monkeys and apes have different limb proportions. Monkeys walk on all fours with their palms down, similar to a dog or a cat, and tend to have front and rear limbs of near equal size, with their spines parallel to the ground. Living apes, on the other hand, have longer arms than legs. *Proconsul* had monkey-like limb proportions, and likely walked on top of branches as do living monkeys, as compared with living apes, who more often hang by their arms under the branches. This does not mean that *Proconsul* was a monkey as it had ape-like teeth. In addition, although the limb proportions of *Proconsul* resemble that of monkeys, their skeletons also show a distinct ape trait—they did not have a tail. Although monkeys have tails, as do other primates, neither apes nor humans have a tail. *Proconsul* looks a bit like an ape, but also a bit like a monkey.

We can see other examples of this mix of monkey and ape characteristics throughout the skeletal anatomy of *Proconsul*. The skull is large relative to body size, more like that of apes than monkeys. The arms and hands are more monkey-like, whereas the shoulders and elbows are more ape-like. Taken all together, this mixture of ape and monkey traits suggests that it is somewhere near the split of old-world monkey and ape lines on the primate family tree. Does this make *Proconsul* an early monkey, an early ape, or the common ancestor of both lines? In evolutionary analysis, we look for the presence of traits that are shared and derived from a common ancestor to help figure out a species' evolutionary status. A good example in the case of *Proconsul* is the fact that it does not have a tail. Tails are commonplace throughout primates (and other mammals), so we consider the presence of a tail a primitive trait inherited by many forms from an ancient ancestor. The absence of a tail, however, is something shared by apes and humans, and is a derived (new) trait. The absence of a tail in *Proconsul* suggests that it lies on the line that had already branched off from the Old World monkeys. When we consider all traits, we see a picture of *Proconsul* as an early representative of the line leading to the common ancestor of African apes and humans. I say "early" because *Proconsul* retained a number of monkey-like traits, which suggests it had not changed too much since the split from the monkey line in Africa.[5]

With changes in the interpretation of both "*Ramapithecus*" and *Proconsul*, the evidence for an early split of African apes and humans 15 to 20 million years ago began to disappear. The other piece of evidence was not from the fossils, but from comparative biochemical and genetic analysis of living primates. The earliest work in this area was done by bacteriologist George Nuttall in the early twentieth century, who examined the reaction of blood proteins across different primate species. The more closely related two species are, the greater the reaction between blood antigens and antibodies. Based on his comparisons, Nuttall found that humans were more similar to other Old World primates than to the New World monkeys. (The descriptive term "Old World" refers to Africa,

Europe, and Asia, whereas "New World" refers to the Americas.) In the 1960s, Morris Goodman performed comprehensive analyses of immunological reactions based on the blood of different primate species. His results showed that humans and the great apes are more similar to each other than either is to monkeys. Further, he showed that African apes and humans were actually more closely related to each other than either is to the Asian great ape, the orangutan. This latter finding runs counter to cursory examination of orangutans, African apes, and humans. After all, both orangutans and African apes walk on all fours, have smaller brains and bigger canine teeth than humans have, and are hairy. Although this quick physical inspection might suggest that African apes and orangutans are more closely related to each other than either is to humans, Goodman's immunological analysis showed that the closer kinship was between the African apes and humans, a result confirmed by all subsequent genetic analyses.[6] It turns out that shared primitive traits of apes (such as hairiness and large canines) and unique features of humans (such as larger brains and walking upright) are not useful in resolving evolutionary relationships. The most accurate reconstructions of a family tree are based on the number of shared derived traits, as these indicate evolutionary relatedness. In addition, genetic data have increasingly been used as a yardstick to measure evolutionary relationships.

By the time Goodman's results were published, scientists were realizing that comparative molecular analysis could provide information on the evolutionary relationships of living species. In other words, biochemical and genetic analyses can help us build a family tree by linking species that are genetically most similar to a more recent common ancestor. In the case of the great apes and humans, these results show that the line leading to the orangutan diverged first, followed later by the divergence of African ape and human lines. Current genetic data can provide even greater specificity, showing that the gorilla line branched off from an African ape line before the split of the human and the chimpanzee-bonobo lines.

The rising field of molecular anthropology underwent a major shift in the 1960s with the development of methods to date the divergence of species. Implications of this work for human evolution came from the work of anthropologist Vincent Sarich and biochemist Allan Wilson, who developed a way not only to reconstruct a family tree showing the relationship of humans and other primate species from molecular data, but also to provide an estimate of *when* the different key events occurred. Sarich and Wilson presented their conclusions in three papers published in 1966 and 1967.[7] The first of these papers outlined an improved method for looking at similarities of different species based on the albumin protein. They found that humans were more similar to African apes (chimpanzee and gorilla) than to other apes (the Asian apes, the gibbon and orangutan) or monkeys. In a footnote for the first paper, they suggested that their measure of immunological distance was a reflection of the time since species split off from a common ancestor and, further, that apes and humans "share far more recent common ancestry than is generally supposed."[8] This idea is followed on in their 1967 paper on an "Immunological Time Scale for Hominid Evolution," which established a method for dating the split of humans and apes. Their basic method was to start with their findings that the albumin protein had evolved at a stable rate in different evolutionary lines, as demonstrated in their other 1967 paper. Consequently, the immunological differences between two species reflected the amount of time since those two species shared a common ancestor. Species that were more closely related (i.e., had a more recent ancestor) were more similar biochemically than species that shared a more distant ancestor. This idea is analogous to your genetic relationship with various relatives. You are more genetically similar to a sibling than a second cousin, as you share more recent common ancestors with your sibling (your parents) than with your second cousin (with whom you share some of your great-grandparents), and so on.

Sarich and Wilson's findings showed that you could express the length of time since two species shared a common ancestor in terms of units of immunological difference. They then reasoned that if they knew the actual geologic date of at least one branch in a family tree, they could then estimate the actual geologic date of the other branches by noting the ratio between immunological difference and geologic age and then using this ratio to fill in other dates. Sarich and Wilson took the estimated date (based on fossils) of 30 million years for the split of the Old World monkeys as their calibration point. Because the immunological difference between humans and the African apes is roughly a sixth of that between hominoids (apes and humans) and Old World monkeys, the inference is that the age of the split of humans and African apes is a sixth of the age of the split of the Old World monkeys. Taking one sixth of 30 million years gives an estimate of the date of the human–African ape split as $1/6 \times 30$ million = 5 million years.

Since the original work by Sarich and Wilson, numerous studies have used molecular data to estimate the split of the chimpanzee and human lines. Over time, the methods became more precise and the available data expanded from indirect assessments of genetic similarities through comparison of proteins to direct analysis of DNA differences. In general, these studies have shown that Sarich and Wilson's original estimate of

5 million years was a bit too recent, but not by much, and most recent studies suggest that the human line split off from the African apes between 6 and 7 million years ago.[9]

These dates are in general agreement with the fossil record. We generally define fossils as hominins if they show evidence of having been bipedal. According to this criterion, the oldest definite hominins belong to the 4.4 million-year-old species *Ardipithecus ramidus* from Ethiopia. An earlier species of *Ardipithecus* dates back close to 6 million years and may be bipedal as well, as might also be the case for a form known as *Orrorin* from Kenya, dating back 6 million years, and *Sahelanthropus* from Chad, dating back perhaps to 7 million years ago. There is still debate over the hominin status of these older forms, but even the oldest is dated close to the molecular estimate.

There may be further adjustments to the genetic estimates. The longer the generation length the larger the estimated age of a split. Estimates of generation length from wild chimps and gorillas suggest longer generation lengths, which in turn gives revised estimates of 7 to 12 million years for the African ape and human split.[10] Our estimates of mutation rates are also subject to change. Some new estimates of mutation rates suggest a somewhat older split of African apes and humans, perhaps between 8 and 10 million years ago.[11] Even so, these revised dates are still much more recent than the estimates of 15 to 20 million years ago that were based on preliminary assessment of *Proconsul*.

Myth #11 *Gigantopithecus* was the ancestor of "Bigfoot" (assuming Bigfoot exists)

Status: An extinct form of ape has been named Gigantopithecus *(= "giant ape") because of the huge size of its jaws and teeth. Some have proposed that* Gigantopithecus *was a biped and a likely ancestor of "Bigfoot," a supposed large bipedal ape-like creature that the majority of the scientific community does not accept because of scant supporting evidence. Even if Bigfoot exists, the supposed link to* Gigantopithecus *does not have strong support, as there is no evidence how* Gigantopithecus *walked and it is not likely that such a large creature could walk on two legs.*

A frequent question about ape and human evolution concerns "Bigfoot," the popular name for a very large bipedal ape that has been suggested to live in forested regions such as the Pacific Northwest region of the United States. A similar animal, the yeti (or "abominable snowman"), has been suggested to live in the Himalayan region of Asia. A casual search on Google shows heated debate between those that adamantly believe in the Big Foot's existence (including those that report eyewitness accounts) and those that cite a lack of credible evidence. Advocates point to footprints, impressions of skin ridges, and a film as evidence that Bigfoot exists. Skeptics point to a lack of any physical remains (such as a skeleton or teeth), frauds, and the suspicion that the film images are of someone dressed in an ape suit. In general, the scientific community has not accepted the existence of Bigfoot,[12] although there are exceptions.[13]

Many anthropologists remain open to the possibility, generally thought to be extremely remote, that Bigfoot might be real, but are not willing to acknowledge its existence based on what has been offered as evidence to date. As anthropologist Matt Cartmill notes, "Many of the supposed Bigfoot traces are clearly hoaxes. Others might be genuine, but none of them is beyond the scope of ingenious trickery. The only way to settle the issue is to show us a specimen."[14] So far, no direct physical evidence has been provided that has convinced more than a few anthropologists that Bigfoot exists. One suggested source of evidence has been hair samples thought to have been from Bigfoot or some other anomalous primate. DNA testing of 30 different samples showed that they all belonged to known mammals, primarily various bear species, but also including hair from horses and cows, among others.[15]

Although such a primate is *possible*, most consider the appropriate null hypothesis to be that Bigfoot does not exist, pending convincing direct evidence in the form of skeletal and/or dental remains to show otherwise. Flipping the null hypothesis around and starting with the idea that Bigfoot exists until proven otherwise makes little sense because, as Cartmill points out, "It's hard to demonstrate that something doesn't exist."[16]

Some of the indirect evidence point to the fossils of *Gigantopithecus*, a very large ape that lived in parts of Asia from about 9 million years ago until several hundred thousand years ago, as support for the existence of Bigfoot. At one level, the fossil evidence for *Gigantopithecus* shows us that there have been apes larger than gorillas and orangutans, which are pretty big to begin with. *Gigantopithecus* is part of a large collection of fossil evidence that shows us that apes as a group were quite diverse in the past, with many different forms other than those that have survived to the present. However, *Gigantopithecus* has been used in the Bigfoot debates as more than an example of potential primate diversity; it has also been claimed by some that *Gigantopithecus* is the ancestor of Bigfoot. (Obviously, those that make this claim assume that Bigfoot is real and therefore needs an ancestor.)

Who was *Gigantopithecus* and how big was it? The oldest fossils of *Gigantopithecus* date to the Miocene epoch, a time in earth's history between 5.3 and 23 million years ago, during which the first apes appeared, including *Proconsul* (discussed in the previous myth). There was a wide range of diversity in apes later in the Miocene, with dozens of different species spread out over Europe, Asia, and Africa. There were many more species of apes living during the Miocene than are around today, showing a general trend toward the extinction of ape species over time. Some Miocene species are similar in some ways to living species, while others are quite different, including *Gigantopithecus*. The oldest remains of *Gigantopithecus* have been found in India and Pakistan and date to about 9 million years ago. It is considered a different (and smaller) species from the remains found in East Asia, including China and Vietnam, that date to several hundred thousand years ago. This relatively recent date means that *Gigantopithecus* lived at the same time as the early human species *Homo erectus*, which suggests that they likely encountered one another. From the available evidence, it looks like *Gigantopithecus* became extinct by 200,000 years ago.[17]

The first discovery of *Gigantopithecus* was a single tooth found by paleontologist Ralph von Koenigswald in an apothecary shop in China, where fossil teeth are often sold as "dragon bones" and used in a variety of folk remedies. Because of the massive size of the tooth, Von Koenigswald named a new genus of fossil ape—*Gigantopithecus*—which translates as "giant ape." Since that time, *Gigantopithecus* fossils have been found in India, China, and Vietnam.

The actual fossil evidence is somewhat limited. To date, we only have dental remains of *Gigantopithecus*, consisting of over one thousand isolated teeth and four lower jaw fragments. The teeth and jaws are massively large, which implies that the overall body size was also large. Using the relationship of jaw and body size in gorillas as a guide, it has been estimated that the larger species of *Gigantopithecus* may has been about 2.75 meters (9 feet) tall and weighed between 150 and 230 kilograms (331 and 527 pounds).[18] Other estimates suggest *Gigantopithecus* may have been even a bit taller, at 3 meters (10 feet), and more massive, up to 545 kilograms (1,200 pounds).[19] Although jaw measures are not the best for estimating body size, the results do suggest a creature even larger than a modern male gorilla. The name of "giant ape" is well earned.

Is *Gigantopithecus* the ancestor of Bigfoot? Obviously, if Bigfoot does *not* exist, the question of its ancestor is moot. Should we examine the possibility of an evolutionary relationship with *Gigantopithecus* under the possibility that Bigfoot *might* exist? In terms of an actual research program, I would argue no, but given the strength of belief one sometimes sees in support of Bigfoot, I think it useful to examine the claims for a relationship with *Gigantopithecus*. For the sake of argument, we will perform a thought experiment assuming Bigfoot's existence to focus on the specific claims made for *Gigantopithecus* ancestry.

The idea that Bigfoot represents either surviving populations of *Gigantopithecus* or a species that is descended from *Gigantopithecus* has been noted by a number of anthropologists at different times, although more as an interesting possibility than as an established fact. If one had to pick a possible Miocene ancestor of Bigfoot, *Gigantopithecus* makes a good choice in terms of dating because it lived until about 200,000 years ago, making any claim that it could have survived until the present more reasonable than fossil apes that have been extinct for many millions of years. *Gigantopithecus* is also large, and advocates of Bigfoot stress a large body size based on their suggested footprint and eyewitness accounts. The fact that *Gigantopithecus* is an Asian ape and Bigfoot is a North American phenomenon is no barrier to an evolutionary relationship, and advocates point to an Asian origin followed by a migration over the Bering Land Bridge that connected northeastern Asia and North America at various times in the past.[20] However, no direct evidence for this migration has been found, as no *Gigantopithecus* fossils have been found in North America. Indeed, no fossil apes or humans other than *Homo sapiens* have been found in the New World.

It has been suggested that there is an evolutionary link because both Bigfoot and *Gigantopithecus* were bipeds. In reality, the idea that *Gigantopithecus* walked on two legs is entirely speculative and not supported by the fossil evidence. We do not have *Gigantopithecus* bones that would show whether they were bipeds or not, which are mostly the pelvis, knees, and leg bones. In fact, we do not have any *Gigantopithecus* fossils other than several lower jaw fragments and isolated teeth. Any inferences about how *Gigantopithecus* have to come from these remains and comparative primate anatomy.

The late anthropologist Grover Krantz claimed that the limited fossil evidence did support the idea that *Gigantopithecus* was a biped. Krantz noted that the jaw of *Gigantopithecus* diverged toward the rear, which meant that the mandibular condyles (the parts of the lower jaw that connects with the skull on both sides) were widely separated. He further proposed that this shape meant that the neck would have been positioned between the sides of the lower jaw, rather than to the rear as in living apes, a position that implies that the skull was positioned on top of the body. Therefore, according to Krantz, *Gigantopithecus* was a biped.

Furthermore, Krantz suggested that Bigfoot was actually a species of *Gigantopithecus*.[21] However, it has not been established that bipedalism can be inferred from jaws and, in any event, the *Gigantopithecus* jaw fragments are missing the back portions and the mandibular condyles.[22] Thus, the evidence for bipedalism is lacking.

Further, it seems unlikely that such a large creature could walk successfully on two legs. It seems more likely that *Gigantopithecus* would have walked on all fours, which would provide more support for its huge size. In addition, because *Gigantopithecus* represents an evolutionary line that split off *before* the split of apes and hominins, for it to have been a biped would mean that the evolutionary changes for bipedalism would have had to occur independently in the hominin and *Gigantopithecus* lines, something that seems rather unlikely.[23]

There have also been suggestions in the past that *Gigantopithecus* was a hominin based on dental evidence and, because hominins are generally defined as bipeds, the presence of hominin dental traits meant that *Gigantopithecus* was also a biped. It is true that *Gigantopithecus* shares certain dental traits with humans and early hominins, but these similarities are no longer taken to indicate kinship. *Gigantopithecus* has thick enamel on its molar teeth, much like hominins, but so do orangutans. Enamel thickness is no longer regarded as a diagnostic hominin trait. *Gigantopithecus* also has relatively small canines, but this too is something that is also not a clear diagnostic as canine size varies quite a bit in fossil apes and some species have reduced canines but are not bipeds. The argument that *Gigantopithecus* was a hominin based on dental traits is not accepted.

Although it remains a remote possibility that Bigfoot exists, the available evidence does not support the hypothesis that *Gigantopithecus* was its ancestor. Noting the *possibility* of an evolutionary connection is sufficient to state a hypothesis, but nothing further can be done without some direct evidence to support this hypothesis. We should remain open to new evidence, but being open to evidence supporting a hypothesis is not the same as accepting the hypothesis without such evidence. From what we can tell, *Gigantopithecus* left no descendants.

Myth #12 Human traits all evolved at the same time

Status: A common image of human evolution is that of a gradual transition from a quadrupedal, small-brained, large-toothed ape into a bipedal, large-brained, small-toothed modern human, where the changes in locomotion, brain size, and tooth size all took place at the same time.

The fossil record now rejects this view, showing instead a picture of a mosaic of changes, with different changes taking place at different times. Upright walking evolved millions of years before a significant increase in brain size.

A common image of human evolution is one that shows a sequence of pictures representing slow, gradual change from an ape-like form into a modern human. This imagery is used quite often in advertising to show the evolution of products from some earlier form into the most current version. A typical image shows an ape-like creature at one end of the sequence walking on all fours, with a small brain, large face and teeth, and a hairy appearance. At the other end of the sequence is a modern human, walking upright with a large brain, small face and teeth, and relatively hairless. In between these images generally lie a series of intermediate steps, such as an ape that is starting to walk on two legs, but is still not entirely upright and having a somewhat larger brain, a somewhat smaller face and teeth, and some loss of hair. Each step between ape and human typically shows the gradual and simultaneous reduction of ape characteristics being replaced by increasingly human characteristics. Although this common picture is accurate in identifying key physical differences between living apes and humans, the idea that there was a gradual and simultaneous transition between ape and human has been proven wrong by the fossil record.

Two related questions emerge when considering the evolutionary relationship of apes and humans. First, what are the differences between apes and humans? Second, when did these differences emerge? The answer to the first question was touched on earlier (Myth 9), and will be expanded on here. The second question will be dealt with here in broad detail using a summary of the key events in human evolution.

First, a bit of review. Genetic and anatomic evidence shows that our closest living relatives are chimpanzees and bonobos, both African apes (the third African ape, the gorilla, is genetically a little bit more different). We compare ourselves to chimps and bonobos to see both our similarities and our differences. Similarities include our mobile shoulder joint (allowing them and us to lift our arms above our heads), lack of a tail (as compared with other primates and mammals), and similar cusp patterns of our molar teeth, among other traits. Such similarities reflect our descent from a common ancestor; for example, neither apes nor humans have a tail because they are both descended from a common ancestor that did not have a tail. Our differences from chimps and bonobos as well as other primates provide clues about what has changed in the course of human evolution. Some human traits are not shared with other primates; these are *unique* traits that evolved at some point since the time when we shared a common ancestor with chimpanzees and bonobos.

What are the most obvious unique human traits? For one thing, we are bipedal—we walk habitually on two legs. Chimps and bonobos (as well as gorillas) move around differently. When they are on the ground, the African apes most frequently knuckle walk, which is a form of quadrupedal (four-footed) locomotion where they rest their front limbs on their knuckles rather than on their palms. However, this is not the only way they move. Chimps and bonobos are also excellent climbers, and also can (and do) walk on two legs at times. Although we describe these apes as knuckle walkers, they are in fact capable of different types of movement on the ground and in the trees. We have lost that flexibility and have become more specialized as bipeds. It is not that we are bipedal and the chimps and bonobos are not; instead, we are better at it. At some point in our evolutionary history, we made the transition from occasional bipedalism to what we call obligate bipedalism. In other words, we *have* to walk on two legs. Other than as infants, we cannot move around on all fours (our legs are too long), and we cannot knuckle walk. We can climb to some extent, but other than childhood play and some forms of exercise, this is not something we do that often, and certainly not as a way to get around in our daily lives! We are excellent bipeds, but that is all we do. Our bipedalism is made possible through anatomical changes in our pelvis, knees, and feet.

The other most obvious physical characteristic of modern humans is our large brains. Although primates in general and apes in particular have large brains relative to body size, ours are even bigger. There is a general tendency in primates relating average brain size to average body size; the larger the species' body size, the larger its brain size. If we exclude humans from this comparison, the relationship between brain size and body size among primates is very strong. (In statistical terms, there is a very high correlation.) What is most interesting is when we consider humans in terms of the general pattern of brain and body size in primates. Here, our brains are roughly three times what we would expect in a primate of our size. If we followed the typical primate pattern, our brain size would be about the size of a chimp's, but instead it is much bigger. At some point in our evolutionary line, brain size increased.

A somewhat less obvious difference between apes and humans (until you get up close) is that humans have very small canine teeth, as discussed in Myth 9 about the evolutionary status of "*Ramapithecus*." We have short stubby canines that actually function more like our incisor teeth than the large canines found in monkeys, apes, and most other mammals. Sometime in our evolution, humans evolved smaller, non-projecting canine teeth.

An important feature of bipedalism, increased brain size, and reduced canine size is that they are all traits that preserve in fossil remains. The key changes for bipedalism can be seen in the skeleton. The larger brain can be inferred from the volume of the skull, even though the brains have long since decomposed. The size of canine teeth can easily be measured. Not all ape–human differences are as easy to look at from fossils. For example, we are relatively less hairy than apes; while we still have hair, it is finer and less dense, making us what zoologist Desmond Morris referred to as "naked apes." Somewhere in the course of human evolution, we became less hairy; this change probably correlated with the increase in our sweat glands as our ancestors moved into the open grasslands. For the moment, we will not concern ourselves with hairlessness or other unique human features of soft body tissues, as we want to focus on those traits that can be seen from skeletal and dental remains.

Before doing so, we also need to consider a behavioral distinction of humans that has entered frequently into discussions of human origins—stone tool technology. While it was once common to consider humans as uniquely different because of our ability to make and use tools (Man, the tool-maker, described in Myth 23), we now know that chimpanzees can fashion and use simple tools, such as sticks for fishing termites out of mounds or rocks used to crack open nuts. Given that apes can make and use such simple tools, it seems reasonable to assume that early human ancestors could do so as well. However, we cannot look for direct evidence of this behavior. If an early human ancestor millions of years ago used a stick to fish for termites, spear an animal in a tree (as chimps have been found to do), or dig in the ground for edible roots, that stick will have long since decomposed, and we would not find it. Stone tools, on the other hand, can preserve for millions of years. Thus, if an early ancestor used one stone to remove some flakes off another stone to produce a sharp edge, we could (and do) find it. The archaeological evidence allows us to date the first appearance of stone tools.

At this point, we have identified three anatomical traits (bipedalism, big brain, and small canines) and one behavioral trait (stone tool technology) that we can examine from the fossil and archaeological records to figure out when they first appeared. Although there is still debate over exactly how old bipedalism is, the evidence to date shows that it evolved well before brain expansion or the first use of stone tools. The oldest suggested evidence of bipedalism comes from the species *Sahelanthropus tchadensis* discovered in Chad. (The name translates as "the man from Chad, in the Sahel"; the Sahel is an area of Africa north of the Saharan desert.) The main evidence for *Sahelanthropus* is a distorted skull that appears to date back as much as 7 million years. The skull has a mix of human-like traits (such as a relatively non-protruding face) and ape-like traits (such as a very small brain size). The reason that *Sahelanthropus* has been suggested to be a biped is the fact that the large hole in the skull where the spinal cord enters, the foramen magnum (Latin for "big hole"), is located underneath the skull. We also have our foramen magnum directly under our skull, and this reflects the fact that we stand on two legs, because our skull sits on top of our spine, which enters the skull at the bottom. In apes, however, the foramen magnum is located farther back on the skull, which reflects the fact that an ape's skull is in front of the body, and not on top. The position of the foramen magnum is thus a clue to whether an organism was bipedal or not. Some scientists have argued that *Sahelanthropus* is a biped based on the position of its foramen magnum; the problem here is that the skull is distorted, which might be give us an inaccurate impression. Digital reconstruction of the skull has been used to correct for the distortion, and the results suggest that *Sahelanthropus* was indeed a biped.[24] If confirmed, this pushes the first appearance of bipedalism back between 6 and 7 million years ago.

Another candidate for an early biped is the species *Orrorin tugenensis* ("Original man from the Tugen Hills") from Kenya and dating back 6 million years ago. This species is poorly known, consisting of dental and skeletal fragments. One of these remains is part of a femur, the upper leg bone. The anatomy of the *Orrorin* femur strongly suggests bipedalism. Another species close in time is *Ardipithecus kadabba* ("*Ardipithecus*" = "ground ape," "*kadabba*" = "oldest ancestor"), found in Ethiopia dating back 5.8 million years ago. Most of the remains for this species are dental and skeletal fragments, but a toe bone suggests that it was bipedal. Both *Orrorin* and *Ardipithecus kadabba* suggest bipedalism is close to 6 million years old.[25]

Although the early fossil evidence is fragmentary, it does suggest that bipedalism goes back at least 6 million years ago (and maybe earlier, in the case of *Sahelanthropus*). However, the data to date are not conclusive. Next up in the fossil record is the species *Ardipithecus ramidus* ("*ramidus*" = "root"). Much of the evidence for bipedalism comes from the amazing reconstruction of a single individual nicknamed "Ardi." This species will be covered in more detail in Myth 14, but for now I note that the fossil shows an early stage of upright walking, where the species still spent a lot of time in the trees, but walked upright on the ground, and had already shown some of the pelvic changes associated with bipedalism. *Ardipithecus ramidus* was found in Ethiopia and dates back 4.4 million years ago. Shortly after this time, we have evidence for another definite biped, the species *Australopithecus anamensis*, which lived in Kenya about 4.2 million years ago ("*Australopithecus*" = "southern ape," after the fact that the first species of this genus was discovered in South Africa, "*anamensis*" = "lake").

Overall, we have evidence of early bipeds in Africa from between about 4 million and perhaps 7 million years ago, depending on which if any of the earliest species you accept as bipedal. For our purpose here, it is sufficient to note that even the minimal estimate of the antiquity of bipedalism (4.2 to 4.4 million years ago) rejects the idea that this trait evolved at the same time as other defining characteristics of humans. The largest discrepancy is seen when comparing the early dates for bipedalism with the earliest known evidence of cranial expansion and stone tool technology.

The earliest stone tools have been found in East Africa and are 3.3 million years old,[26] over a million years after even our most conservative estimate of the earliest bipedalism. Cranial expansion begins to take off a little later, at about 2 to 2.5 million years ago. We estimate brain size by calculating the interior volume of the skull using a variety of methods ranging from the relatively low-tech but useful method of seeing how much volume of water or mustard seed is needed to fill the skull, to high-tech measures such as CT scans.[27] Before about 2 million years ago, cranial capacity of early hominins ranged mostly between about 400 and 500 cc (cubic centimeters), which is roughly ape-size and much smaller than a modern human average of about 1350 cc. After 2 million years ago, fossils in the genus *Homo* show a major increase in brain size.[28] These data show clearly that cranial expansion, as with stone tool manufacture, occurred well after the origin of bipedalism.

Finally, we consider the loss of the large protruding canine of apes during human evolution. The canine teeth of the two species of *Ardipithecus* show that canine size and the honing complex were reduced sometime between 6 and 4.4 million years ago. Later hominin species belonging to the genus *Australopithecus* show further reduction in canine size over the next 2 million years.[29] Although it was once suggested that tool use had led to the reduction in canine size, because stone tools would replace the function of larger canines, it is clear from the fossil record that canine reduction began before stone tools and larger brains.

The idea of a simultaneous transition from ape to human involving coordinated changes in upright walking, canine reduction, increased brain size, and the origin of stone tool technology is an appealing model in many ways because it seems to tie everything together in one nice neat package. However, it is wrong. Bipedalism and canine reduction took place long before larger brains and stone tools, and their origins were for different reasons—the topic of later myths.

Myth #13 Large brains evolved very early in human evolution

Status: It was once thought that because of its complexity, the human brain must have taken a very long time to evolve, which in turn suggested that large brains appeared early in human evolution and before other human traits. This view was once supported by a fossil known as Piltdown Man, which appeared to have the large brain of a modern human and the jaws and teeth of an ape. It turns out that Piltdown Man was a hoax. Although the fossil is a fake, the history of the Piltdown hoax provides us with an excellent example of the corrective nature of science.

Of all the changes that have taken place in the course of human evolution, the evolution of our large and complex brains is perhaps the most fascinating. It is the brain that we think about when contemplating the daily lives of our ancestors and their thoughts. Our brains are three times the size expected for a primate of similar body size and have evolved to handle abstract thought, complex problem-solving, language, and culture. Apes are certainly intelligent, can make and use simple tools, have culture, and can (under the appropriate circumstances) learn rudimentary aspects of language (see Myth 23). However, there is also an obvious gap between their cognitive abilities and ours, as no one expects apes to equal our grasp of technology, or to discuss current events or the nature of reality! One of the goals in the study of human evolution is to document, where possible, the evolution of the human brain and the evolution of our current abilities from an ape-like beginning.

At the turn of the twentieth century, there were few fossil remains of human ancestors to look at when considering the evolution of human characteristics such as the brain. Scholars debated the sequence of events in the evolution of human traits—did bipedalism or larger brains come first? On the other hand, did bipedalism and large brains come about at the same time? As shown in the previous myth, the fossil evidence today is complete enough to answer this question and shows that bipedalism evolved a long time before a major expansion of the brain.

Yet, at the beginning of the twentieth century, the fossil evidence was much more limited and the view that large brains evolved very early in human evolution was accepted by a number of scientists. Given the complexity of the human brain, some considered that it would have taken a very long time to evolve from an ape-like brain to a human brain, which in turn meant that large brains had to have been around for quite some time. Those who advocated this "brain first" idea of human evolution were therefore very pleased when their views were seemingly vindicated by the discovery of a fossil that appeared to represent a species with a large modern brain case and the jaw of an ape! This fossil showed that large brains evolved before human characteristics of the jaws and teeth.

This discovery took place in the early twentieth century in the village of Piltdown, located in East Sussex County in England. At a scientific meeting in 1912, lawyer and amateur archaeologist Charles Dawson announced the results of several excavations at a gravel pit in Piltdown. According to Dawson, in 1908 he had been given the remains of a skull by a worker at a gravel pit in Piltdown. Dawson's subsequent excavations revealed jaw fragments, stone tools, and a number of other mammal bones that tentatively dated the finds to the Early Pleistocene, a geologic time now dated between 780,000 and 2.6 million years ago. At the time, accurate methods of dating geologic times had not been invented, so they did not know exactly how old the Piltdown remains were, other than being ancient.

The skull fragments were consistent with a large-brained modern human, with a brain size only slightly smaller than that of an average living human. The jaw fragment consisted of part of the lower jaw (the front was missing) with two molar teeth. The back part of the jaw, which attaches to the skull, was also missing. The overall shape of the jaw was definitely ape-like, although the molar teeth were worn flat in the same manner as a human.

The cranial and dental fragments appeared to be of the same age because they showed similar amounts of discoloration, which also supported the idea that the skull and jaw fragments came from the same individual. Thus, it appeared to some authorities that the Piltdown remains were evidence of a missing link between apes and humans—a primate with the large brain of a human but the jaw of an ape. The Piltdown Man specimen supported the idea that one of the first changes in human evolution was the development of a large brain. Further, Piltdown Man appeared to be very ancient, which supported the idea that a structure as complex as the human brain must have taken a long time to evolve. Sir Arthur Smith Woodward, of the Department of Geology at the Natural History Museum in London, assigned Piltdown Man to a new species, "*Eoanthropus dawsoni*," which translates as "Dawson's dawn man," named after the discoverer.

Not everyone accepted the conclusion that Piltdown Man represented an ancestor of modern humans that had a large brain and an ape jaw. Some questioned whether the cranial and dental remains actually belonged to the same individual. Parts of the skull and the back of the jaw that might have disproven an association were (conveniently) missing. Others questioned an initial reconstruction that the front of the jaw was ape-like, as that part of the fossil was also missing. However, subsequent excavation (conveniently) turned up a canine tooth resembling that of a chimpanzee, providing support for the view that the front of the jaw was ape-like.

Although there were those who questioned the assumption that the human-like skull fragments and ape-like jaw fragments belonged to the same individual, many viewed Piltdown Man as an acceptable ancestor because it fit in with their preconceived notions of the antiquity of a large human brain. In any event, it is clear that Piltdown Man (and the belief of the large human brain evolving first) affected reactions to the later discovery of other fossils far away in South Africa. In 1924, a fossil found at the limestone quarry at Taung in South Africa was brought to the attention of Raymond Dart, a young anatomist at the University of Witwatersrand in Johannesburg. The fossil was embedded in stone, and with great patience Dart was able to uncover a facial fragment including the upper jaw, the mandible (lower jaw), and a complete endocast (a cast of the interior of the skull). These remains belonged to a three-year-old primate with an interesting mix of ape-like and human-like traits (Figure 2.3). Although the brain size was small and similar to that of an ape, the shape of the jaw and the small canine teeth were human traits. Further, Dart was able to argue that the foramen magnum was located closer to the center of the bottom of the skull than in apes, thus showing that this individual was a biped. In a famous paper in 1925, Dart named a new species based on this find—*Australopithecus africanus*, which translates as the "southern ape from Africa." (Many early hominin species have since been assigned to the genus *Australopithecus*, but this was the first.)

Although scientists at the time thought the Taung child (as the specimen was known) was interesting, there was little support for Dart's views that *Australopithecus africanus* was a human ancestor. Most felt it was some other type of primate with no direct relevance for human evolution. Part of the reason that Dart's ideas were not accepted was the fact that the Taung specimen was a young child, and species characteristics are

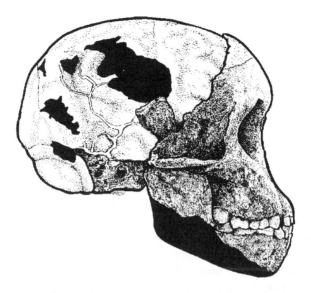

Figure 2.3 The Taung child, the first discovered specimen of the species *Australopithecus africanus*. This specimen consisted of a face with upper jaw, a lower jaw, and an endocast, which is a mold of the interior of the brain case, providing information about the shape of the skull and some clues regarding brain anatomy. Note that the Taung child has the small brain and large face of an ape and human-like jaw and teeth, just the opposite of Piltdown Man. Source: Cartmill and Smith, 2009. Reproduced with permission of Wiley and Matt Cartmill.

often difficult to assess in youngsters. However, it is generally conceded that many rejected Dart's evidence in part because it totally contradicted Piltdown Man. The Taung child had a small ape-sized brain with human-like jaws and teeth, just the opposite of the large human brain and ape jaw seen in Piltdown Man. As these two specimens were completely at odds with one another, one could only accept one of them as a human ancestor. Given the prevalence of the idea of a large and ancient human brain, Piltdown Man was the clear favorite.

Not everyone rejected Dart's idea. One prominent supporter was the famed paleontologist Robert Broom, who had had a notable career that included discovery of many mammal-like reptiles, a group of early reptiles that were transitional to mammals. Broom also had a reputation for having been eccentric, including excavating in the nude and having dropped to his knees in admiration when he first saw the Taung fossil. In any event, Broom spent a number of years at various sites in South Africa where he found several species of early hominin, including *adult* specimens of *Australopithecus* that showed that Dart was right. The early stages of human evolution consisted of smaller-brained bipeds. Ultimately, enough fossil evidence had accumulated to reject both the "brain first" model of human evolution as well as Piltdown Man. Over time, the case for *Australopithecus* as an early human ancestor became stronger and Piltdown Man was pushed off to the side as something that did not fit with other, more compelling, evidence. The idea that the cranial bones and jaw fragment belonged to different creatures fit the fossil evidence better.

However, Piltdown turned out to be more than a case of mistakenly assuming that bones of a human and an ape belonged to the same individual. Instead, detailed analyses done by Joseph Weiner and colleagues in the early 1950s showed that Piltdown Man was a fake! Their initial work showed that the jaw had been broken and the molar teeth had been filed down. They were later able to demonstrate that the cranial bones and jaw were not of the same age using a method known as fluorine dating. Fossils in the ground absorb fluorine from groundwater over time, and the longer a bone has been in the ground, the more fluorine will be absorbed. Because the amount of fluorine absorption varies from place to place, this method cannot be used to give a direct answer of how old a fossil is, but it can be used to compare fossils at the same site and determine if they were of the same geologic age. In the case of Piltdown Man, the fluorine test showed that the cranial bones and jaw fragment were not the same age, and therefore could not belong to the same individual. Further, comparison of fluorine content in the Piltdown Man specimens with other bones at the site that were typical of animals found in the Early Pleistocene showed the Piltdown remains were not that old. Eventually it was shown that the jaw fragment belonged to a present-day orangutan and the cranial remains were from a slightly older modern human. Both had been modified, including filing the ape teeth to make the molars look human-like and breaking the jaw so that the parts that would show a lack of fit were missing. All of the remains were chemically stained to make them look ancient and of the same age. Further research showed that the canine tooth came from a modern chimpanzee and had been altered.[30]

Although we have known for many decades that Piltdown Man was a hoax, we still do not know who perpetuated the hoax. Much discussion has gone into trying to solve the mystery of "whodunit?" There are a large number of potential suspects linked to Piltdown Man that had motive, means, and opportunity. Alas, the evidence is circumstantial.

Among the list of suspects is the discoverer Charles Dawson, several leading scientists of the time (including Sir Arthur Woodward, who named Piltdown), the Jesuit priest and paleontologist, Teilhard de Chardin, and Sir Arthur Conan Doyle, the creator of Sherlock Holmes. Archaeologist Kenneth Feder has reviewed the cases for and against these (and other) suspects in the Piltdown affair, and concludes that the case against many of them is weak, but there is compelling (though not definitive) evidence implicating Dawson. It is also possible that more than one person was involved in the hoax.[31]

There are two lessons we can draw from the Piltdown affair, one pessimistic and one optimistic. The pessimistic view is to question the ability of those scientists that were fooled by the Piltdown remains, particularly as certain aspects of the forgery were crude and in retrospect should have been noticed. Although questions were raised about Piltdown pretty much from the start, it took many years to convince the scientific community. The usual explanation is that people see what they want to see, and the prior acceptance of a "brain first" model of human evolution biased people to focus on what evidence fit their views. We are all guilty of selective reasoning at one time or another and the Piltdown affair is an example of this type of bias.

However, we should not take this case as proof of a view that scientists cannot be trusted because they get things wrong. As noted in the introduction to this book, much of the scientific method consists of getting things wrong and then correcting them. A hallmark of science is its openness and commitment to reexamination and the ability to reject a hypothesis when it is no longer supported by evidence. After all, the fact that you are reading about this hoax shows that it did not hold up over time. It was not as though Dawson and others proclaimed that Piltdown Man was our ancestor and all further discussion stopped because of their assumed infallibility. Results are *always* questioned and reexamined, regardless of who states them. The hypothesis that Piltdown Man was our ancestor is not tested by power of authority—we do not accept this view simply because Dawson or Woodward said it was so. Instead, Piltdown was questioned repeatedly and constantly reevaluated in light of new evidence, such as the discovery of *Australopithecus*, and ultimately found not to fit. The further evidence that it was not just a mix of ape and human fossils, but instead was a fake, was the final nail in the coffin for the Piltdown hypothesis. Yes, it is unfortunate that some scientists were fooled for a long time, but in the long run, the self-correcting nature of science won out.

Myth #14 The common ancestor of African apes and humans walked like a chimpanzee

Status: The similarities of African apes and humans show that both evolved from a common ancestor. Because the African apes are all knuckle walkers, some models of human evolution have proposed that this common ancestor was also a knuckle walker. The analysis of a partial skeleton of the species Ardipithecus ramidus, dating back 4.4 million years ago, suggests that the earliest hominins differed from chimpanzees in a number of ways, and that the last common ancestor might not be a knuckle walker.

A common misconception about human evolution is that we evolved from a *modern* African ape, such as a chimpanzee or bonobo. As noted in several earlier myths, we did not evolve from modern chimpanzees (or any modern primate), but instead both chimps and humans evolved from a common ancestor, estimated from genetic analysis to have lived about 6 to 7 million years ago in Africa. What this means is that both the human and chimpanzee-bonobo lines have evolved from this common ancestor. As noted in earlier myths, we need to separate out primitive and derived traits in fossil and living creatures to determine exactly what changed, and when. This consideration of comparative anatomy allows us to propose different hypotheses about the pattern of evolutionary relationships between African apes and humans.

The three living African apes (gorilla, chimpanzee, and bonobo) vary a lot in terms of size, ecology, and social structure, but they also share a number of characteristics, including knuckle walking. The arms and legs of monkeys are of similar length and they walk palm down, as do many other four-legged creatures such as dogs and cats. The African apes are built differently; they have longer arms than legs, such that their head are higher than their rumps when walking on all fours. African apes also have a number of anatomical changes in their wrists and fingers that allow them to withstand the stress of putting weight on their knuckles when moving about.

We need to consider the evolutionary implications of knuckle walking. As shown earlier (Figure 2.2 in Myth 9), genetic evidence provides us with an accurate picture of the evolution of the African apes. There was an initial split from a common ancestor about 7 million years ago, resulting in one line leading to gorillas and the other line leading to the common ancestor of chimpanzees and bonobos, two species that split roughly 2.5 million years ago. Let us examine the ancestry of the three African apes in terms of knuckle walking. The two most related apes, the chimpanzee and the bonobo, are both knuckle walkers, which implies that their common ancestor was also a knuckle walker. Further, the genetic family tree shows that the chimpanzee-bonobo line had a common ancestor with the gorilla, who is also a knuckle walker, implying that the common ancestor of all African apes was also a knuckle walker (common ancestor "A" in Figure 2.2). We make these predictions based on the principle of parsimony, which means that it is more likely that two species share a derived trait due to common ancestry rather than an independent origin.

If we follow the principle of parsimony, we can get an idea of what the ancestor of the hominin line might have looked like. Remember that the hominin line split from the African apes *after* the split of the gorilla line. If we assume that all the African apes as far back as their common ancestor were knuckle walkers, then we can infer that the hominin line is descended from a knuckle walker. Because hominins are defined as bipeds, this means that bipedalism evolved from knuckle walking. As such, many have proposed evolutionary models where the common ancestor of humans and chimps (and bonobos) was much like a modern chimpanzee—a knuckle walker with larger canines and a smaller brain. More extensive debate has often centered on which small-bodied African ape is the best model for our common ancestor—the chimpanzee or the bonobo.

What if the common ancestor of the African ape and human lines was *not* a knuckle walker, but moved around differently on all fours, such as does an orangutan or a monkey? If common ancestor "A" in Figure 2.2 was not a knuckle walker, then knuckle walking had to have evolved twice—once in the line leading to the gorilla and once in the line leading to the chimpanzee and bonobo. Ultimately, we have to make a decision about the reason two species (such as gorillas and chimps) share a derived trait—was it because of shared ancestry (a pattern known as homology) or because of independent evolution (a pattern known as homoplasy)?

Although we may prefer common ancestry as the most parsimonious explanation of why all three living African ape species are knuckle walkers, we have to understand that homoplasy does happen in evolution, perhaps more frequently than once thought. A preference for parsimonious explanations does not necessarily mean that parsimonious explanations are correct. We look at the anatomy of the first hominins for clues that could tell us which hypothesis is more likely. As noted in Myth 12, we now have fossil remains of perhaps three very early hominins—*Sahelanthropus*, *Orrorin*, and *Ardipithecus*. Do any of these provide information that suggests a knuckle-walking ancestry? Unfortunately, we do not have any postcranial (below the neck) remains of *Sahelanthropus*, and the postcranial remains of *Orrorin* are too fragmentary to shed much light on this question.

The postcranial evidence from the species *Ardipithecus ramidus* tells us a lot more. Many of the remains of *Ardipithecus* that have been found by paleoanthropologist Tim White and colleagues since the mid-1990s were fragmentary, but included in these remains was a partial skeleton dating to 4.4 million years ago. Because of the fragile nature of the remains and distortion, it took 15 years for scientists to uncover and reconstruct the skeleton, a process that used traditional anatomical methods as well as high-tech medical imaging and virtual reconstruction on a computer. The results were worth the wait; in the October 2, 2009 issue of the journal *Science*, an international team presented a series of papers that examined *Ar. ramidus*, particularly with a focus on the reconstructed partial skeleton ("*Ar.*" is the abbreviation for the genus name *Ardipithecus*). This skeleton, nicknamed "Ardi," was that of an adult female who was about 1.2 meters (roughly 4 feet) tall and weighed about 50 kilograms (110 pounds).[32]

Ardi shows an interesting mix of both primitive (ape-like) and derived (human-like) traits, many of which were quite surprising. Like other early hominins, she had a small ape-sized brain and a protruding face. Her canine teeth were larger than in modern humans, but smaller than in the African apes. We classify Ardi as a hominin because of the skeletal evidence that she was a biped, including a human-like shape of the upper pelvis (broad and short) and a foramen magnum at the bottom of the skull. However, some features of the bottom half of the pelvis are more similar to that of an ape, suggesting that the muscles attached to the pelvis were being used somewhat differently than either a modern human or ape. Like apes, Ardi's big toe was divergent, sticking out much like the thumb on the hand, and unlike the big toe of a modern human, which is parallel to the other toes. However, her other toes were less flexible than in apes, thus providing some of the rigid support needed in a bipedal foot. The limb proportions are very interesting. Instead of longer arms as in apes, or longer legs as in modern humans, the arms and legs are roughly similar in length, which are proportions more like that of a monkey. Compared with us, Ardi's arms hang down closer to her knees.

The pelvis and skull show that Ardi was a biped, but what kind of biped was she? Her anatomy is sufficiently different to show that although she walked on two legs when on the ground, her bipedalism was not the same as ours. Owen Lovejoy and colleagues[33] considered all of these clues and concluded that *Ar. ramidus* was a *facultative* biped, which meant that it was not a biped all of the time. In contrast, modern humans are obligate bipeds; we have to walk upright all of the time. For humans, walking upright is the only game in town and our anatomy reflects a number of evolutionary changes that has allowed us to become a very specialized biped. Lovejoy and colleagues argue that at the time of *Ardipithecus* this transition had not yet happened. Ardi did walk on two legs when on the ground, but also spent a lot of time in the trees, where she would have walked on top of branches much like a monkey. *Ardipithecus* represents a stage where the pelvis had changed to allow a combination of climbing and bipedalism. Later in hominin evolution, further changes would lead to more efficient walking and running as these ancestor became more and more adapted to living on the ground, leading ultimately to a fully committed (obligate) biped while losing climbing ability.[34] Thus, Ardi gives us an excellent window on the evolution of bipedalism.

Further, Ardi allows us to make inferences about the common ancestor of humans and African apes. First, analysis of Ardi's anatomy shows no evidence of any indication of a knuckle-walking ancestry. For example, she does not have the rigid wrist or long metacarpals (part of the hands) expected in an early biped descended from a knuckle walker. Instead, she shows the anatomy consistent with walking on all fours above the branches. Although she was not a knuckle walker, she also does not have the traits expected if she descended from a suspensory climber, something also found in living apes. Instead, Lovejoy and colleagues suggest that the last common ancestor of humans and African apes was likely an above-branch quadruped, unlike any living ape. They propose that the African apes and hominins evolved along different paths from such a common ancestor. The hominins evolved bipedalism, facultative at first and obligate later on. The different African apes then evolved suspensory climbing when in the trees and knuckle walking on the ground.[35]

If confirmed, there are two implications from these findings. First, knuckle walking evolved in parallel in gorilla and the chimpanzee-bonobo lines. Although this may not be the most parsimonious explanation, nature is not always parsimonious. There is increasing evidence that some anatomical changes evolve independently in related lines, and knuckle walking in the African apes looks to be yet another example. Second, because the common ancestor likely moved about quite differently than an African ape either in the trees or on the ground, the living African apes may not provide a useful model for understanding the evolution of human bipedalism. It does not look like our ancestors went through a knuckle-walking stage. Lovejoy and colleagues summarize this very nicely: "The specialized locomotor anatomies and behaviors of chimpanzees and gorillas therefore constitute poor models for the origin and evolution of human bipedality."[36] The reconstruction of Ardi has only been available for a few years, and further analysis and debate is likely to occur. For the moment, Ardi's anatomy supports the view that our common ancestor did not walk like a chimp.

Myth #15 Bipedalism first evolved on the African grasslands

Status: A long-standing idea in human evolution is that the earliest hominins developed bipedalism as a response to the demands of living in a savanna environment, consisting of open grasslands with scattered trees. However, reconstruction of ancient environments shows that the earliest hominins such as Ardipithecus ramidus *lived in woodland environments. Bipedalism had already appeared by the time our ancestors spread into savanna environments. Although living on the savanna most likely influenced the further evolution of bipedalism, it does not appear responsible for its initial origin.*

We have evidence of possible bipedalism in forms such as *Sahelanthropus*, *Orrorin*, and *Ardipithecus kadabba*, which date back 6 million years ago or more. The previous myth outlined the fossil evidence for a bipedal ancestor (*Ardipithecus ramidus*) dating back 4.4 million years ago. Additional evidence for early bipedalism comes from slightly younger species in Africa, including *Australopithecus anamensis*, dating back 3.9 to 4.2 million years ago, and *Australopithecus afarensis*, dating back 3.0 to 3.7 million years ago. As noted in Myth 12, the appearance of bipedalism predates other human traits, such as a large brain and stone tool technology. We have seen how *Ardipithecus ramidus* was a primitive biped that still likely spent a lot of time in the trees, walking upright when on the ground. The big question is why this happened. Why walk upright?

The origin of bipedalism has been a hot topic since Darwin's day. Apes can walk upright, but not as good as humans can. It is easy to see that under certain conditions natural selection would favor anatomical variations that allowed easier and more efficient bipedalism. We can see some of those initial changes in *Ardipithecus ramidus*. The underlying question is figuring out what those conditions were that would have favored bipedalism on the ground.

Many hypotheses have been proposed to explain the evolutionary benefit of bipedalism. Darwin and others favored the view that bipedalism frees the hands to carry tools. Although the fossil and archaeological records show that this would not have been the case, because stone tools appear such a long time after the first bipeds, there are other advantages of having your hands free. For example, you can carry food, which facilitates food sharing—bringing food back to others in your group. You can also carry babies, which facilitates their care and survival.[37] Another advantage of walking upright is that it is energy efficient, meaning that you use less energy when walking. An interesting set of experiments showed that bipedal humans use less energy than chimpanzees knuckle walking at normal speeds,[38] an advantage that would come into play in cases where an organism had to range farther in search of food. There are other ideas as well, to be discussed below. Keep in mind that we should not focus on one single cause as a shift to bipedalism could involve many advantages at the same time.

A common feature of models for the origin of bipedalism that were developed throughout much of the twentieth century was the specific role of adapting to the demands of living on the African savanna. Savannas are open grasslands with scattered trees. Today, savannas make up much of the landmass of central Africa. Many of the ideas that have been proposed for the origin of bipedalism make sense in terms of adaptation to a savanna environment that is more open, where food is more widely scattered, and where escape from predators is more difficult because there are fewer trees. The savannas in Africa appeared over millions of years as climate change led to the rainforests shrinking, first to woodlands and then to more open grasslands. If we assume that the ancestors of hominins lived in such areas, we can expect that they would have faced the need to adapt to this changing landscape.

As the forests shrank, our ancestors would have needed to walk farther in search of food. Under this circumstance, bipedal walking would be adaptive because it is more energy efficient, and the farther you would need to range for food, the more important energy efficiency becomes. Having hands free would be useful for foraging and bringing food back to others in your social group. In addition, by standing upright, you could see farther into the distance over the grass, giving an advantage in avoiding predators, important because there would be fewer trees in which to escape. You would also have your hands free to carry weapons. Although we know that bipedalism came a long time before stone tools and manufactured weapons, a group of early bipeds could still carry sticks and stones with which to scare off predators. Finally, there is an advantage in walking upright on the open grassland in terms of adapting to the heat. If you have to forage on the savanna, you do not have the shade that you would get from the closed canopy of the forest, and it would get very hot moving around in the middle of the day. Experiments have shown that bipeds experience less heat stress during the hottest time of day than quadrupeds because the vertical orientation of the body reduces exposed surface area.[39] Taking all of these factors into consideration makes a strong case for the adaptive benefit of bipedalism in a savanna environment.

Not everyone accepted the savanna origin model. One of the main problems with this model is that it assumes that the first bipeds were evolving in a savanna environment. Although this appeared to be the case when our knowledge of early bipedalism only went back 2 to 3 million years ago, we now know that bipedalism has been around at least 4.4 million years and likely even earlier. What was the environment like in the past? We cannot extrapolate from the present range of environmental conditions. For example, the area of Ethiopia where Ardi was discovered is today extremely dry, hot, and inhospitable. How was it different in the past?

We can reconstruct ancient environmental conditions from a number of clues. Using *Ardipithecus ramidus* as an example, we can look at what the environment was like for Ardi and her kin. Some clues come from the analysis of *Ardipithecus* teeth. Different diets result in different wear patterns on the teeth, which can be investigated under powerful microscopes and compared with mammals whose diet is known. The analyses of *Ardipithecus* teeth do not show a strong preference for eating fruit or hard objects, a pattern seen in later hominins, but instead suggest a diverse omnivorous diet as would be found in a wooded environment. In addition, several *Ardipithecus* teeth were subjected to stable (nonradioactive) isotope analysis, which provides information on the relative abundance of different carbon isotopes, and in turn a general view of an individual's diet. The isotope analysis of *Ardipithecus* teeth shows a diet more typical of a woodland rather than grassland environment.[40]

Further evidence is provided from the fossils of other species, both plant and animal. At the Aramis site (where Ardi was discovered), the field team also recovered over 150,000 fossils of ancient plants and animals, all from strata representing a short interval of time roughly between 100 and 10,000 years in duration. The diversity of species in this short geological instant gives us an excellent view of the ancient environment. For example, a wide range of bird fossils was discovered, and those species that are typically found in open country were rare, whereas the presence of other bird species was indicative of a woodland environment. This is also true for the fossils of small mammals.[41] The larger mammals included leaf-eating monkeys that typically live in the trees and not in open environments. Isotope analysis of the large mammals also points to a closed environment ranging of forests and woodlands.[42]

Overall, the evidence is quite compelling—*Ardipithecus ramidus* lived primarily in a woodland environment and not the open grasslands. A woodland environment is more open than a high canopy forest, but not as open as the grasslands. This type of environment makes perfect sense when considering that the anatomy of *Ardipithecus* shows a mix of tree climbing and terrestrial bipedalism, as a woodland habitat will consist of a mix of trees and areas that are more open. It seems likely that *Ardipithecus* climbed and walked above branches in the trees, and was a biped when on the ground. The take-home message from these analyses is that *Ardipithecus* did not live in the open savanna, and bipedalism more likely began in a woodland environment. The savanna model for the origin of bipedalism does not fit the available evidence. This is also true of the environment of later hominins, such as *Australopithecus afarensis*, which also lived primarily in a wooded forest or woodland environment.

Given this, we need to readdress some of the suggested hypotheses for the origin of bipedalism. Some previous explanations, such as being able to see above the savanna grass or adapting to the hot conditions of the noontime sun on the savanna, clearly cannot be used to explain the *initial* origin of bipedalism because these early hominins had not yet ventured out into the savanna. Visual detection of predators and adaptation to heat stress probably were important to later bipeds once they moved into the savanna, but are not convincing evidence for the initial origin of bipedalism. Other potential advantages of bipedalism could still fit a woodland environment. Carrying food and babies would be adaptive when the forests began to shrink but had not yet become open grasslands. The same could be argued for an idea of bipedalism leading to increased energy efficiency, as early hominins would potentially be on the ground more often in a woodland habitat. Other factors might also come into play in a woodland environment. For example, observations of chimps show that they are most often bipedal when standing up to forage for food when on the ground and when standing on branches in trees.[43]

Our best bet for understanding the origin and evolution of bipedalism is to avoid looking for a single factor and a simple "X causes Y" type of model. Bipedalism is something that would affect all aspects of an organism's life, ranging from feeding to survival, and would lead to a number of advantages in a particular set of environmental conditions. Of course, there are also clear disadvantages of being a biped. We are not as fast and not as stable on two legs compared with four legs. Walking upright takes a toll on our bodies, leading to back pain, flat feet, our knees wearing out, and hernias, all problems resulting from the adaptation of a quadruped body for walking on two legs. Another problem is that the modification of the pelvis for walking upright makes birth more difficult, particularly as brain size increased. These are all good examples of what anthropologist Wilton Krogman referred to as the "scars of human evolution,"[44] indications of successful, yet not perfect, evolution. To put it in other terms, nothing is free in evolution. Often, there are negative consequences associated with adaptive changes. The fact that hominins did evolve to be bipedal shows that the net advantages outweighed the disadvantages.

We can envision a situation where a distant ancestor walked on two legs in certain circumstances, such as foraging while on the ground. As the forests began to shrink, this behavior would become more common as the distance between trees increased, perhaps because of increased energy efficiency. In a woodland environment such as this, a biped such as *Ardipithecus* did quite nicely because of a mixed adaptation to living part of the time in the trees and part of the time on the ground. As the woodlands opened up even further, bipedalism would become more and more common, particularly given other advantages such as food sharing. By a certain point in time, our ancestors made the shift from being facultative bipeds to become obligate bipeds, along with further shifts in anatomy, such as the further modification of the pelvis and the loss of the divergent big toe. By the time our ancestors ventured on to the savanna, further advantages would accrue, including handling heat stress. Later on, there were even further advantages, such as the ability to hold tools and weapons. Whether such a model holds up under further analysis remains to be seen. What is clear, however, is that the initial evolution of bipedalism did not likely take place on the savanna.

Myth #16 Lucy was so small because she was a child

Status: "Lucy" is the nickname given to a partial skeleton of the species Australopithecus afarensis. *When people first see her bones, they realize how small she was and sometimes think that she must have been a child. Methods of comparative anatomy allow us to estimate the age at death for a hominin specimen. In this case, we know that Lucy was a young adult. Her species was relatively small and she may have been smaller because she was a female in a species where there may have been large body size differences between the sexes.*

One of the most famous fossil specimens is "Lucy," the nickname given to a partial skeleton found at the Hadar site in the Afar region of Ethiopia and dating to 3.2 million years ago. Lucy represents one of many specimens found at this site during fieldwork by Donald Johanson and colleagues in the early 1970s. Although the specimen's official identification number is AL 288-1 (for specimen 1 from location 288 in the Afar Locality), she is best known as "Lucy," owing to the field team naming her after the Beatle's song "Lucy in the Sky with Diamonds."[45] What makes Lucy particularly interesting is the completeness of her skeleton, which includes cranial fragments, the lower jaw, vertebrae, and ribs, most of two arms, over half the pelvis, and parts of the legs (Figure 2.4). It is much more common to find fragments of individuals, such as a skull from one or a jaw or leg bone from another, than to find a relatively complete individual. Partial skeletons such as Lucy provide a more complete view of the entire organism, allowing us to examine proportions of different parts of the body, such as comparing the length of arms and legs.

Lucy is recognized as belonging to the species *Australopithecus afarensis*. The species name indicates the Afar region of Ethiopia where the fossils were discovered. Johanson and colleagues placed this species into the genus *Australopithecus* because it was similar in some ways to the species *Australopithecus africanus* that had been named by Dart (see Myth 13), but sufficiently different to warrant its own species. Assignment of different species to the same genus is usually made when these species share a similar adaptive niche. In this case, placing the species *afarensis* into the genus *Australopithecus* makes perfect evolutionary sense, but results in a somewhat confusing translation of the entire name—"The southern ape from the Afar"—because Afar is in *East* Africa, not South Africa. However, the original genus name (*Australopithecus*, referring to South Africa) is used because of the rule of priority when naming species. The genus name *Australopithecus* was named before any related forms had been found in East Africa.

Remains of *Australopithecus afarensis* have been found at several sites in East Africa dating from about 3.0 to 3.7 million years ago, a time that comes after *Ardipithecus* and *Australopithecus anamensis* but before *Australopithecus africanus*. Overall, we see the same general anatomy as in earlier hominins—a small-brained biped. However, *Au. afarensis* is also different from what came before. ("*Au.*" is the abbreviation for *Australopithecus*.) The lower pelvis does not show the ape-like features seen in *Ardipithecus*, but is more similar to modern humans. The big toe does not stick out but is instead in line with the other toes. The limb

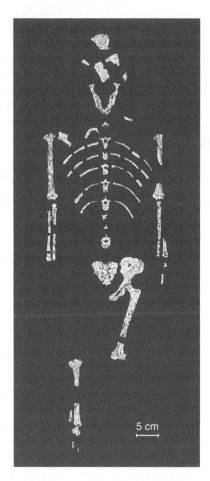

Figure 2.4 Lucy, a partial skeleton of *Australopithecus afarensis*. Source: Cartmill and Smith, 2009. Reproduced with permission of Wiley and Matt Cartmill.

proportions are no longer monkey-like, but the arms are still relatively long compared with modern humans. It appears that bipedalism had further evolved by this time, although Lucy and her kin retained some ape-like traits associated with climbing, such as a more cone-shaped thorax, relatively long arms, and slightly curved toe bones. There has been debate over whether *Au. afarensis* still spent time in the trees or if it was a full-time biped with some ape-like evolutionary "leftovers" in its anatomy. The canine teeth had reduced even more by Lucy's time, but were still large relative to modern humans. The species is interesting because of the mixture of human-like and ape-like traits we see in early hominin evolution. What makes Lucy even more interesting is that you can see enough of her skeleton to get a good feel about how everything fit together, a more thorough picture than one gets from a single bone.

I find that a common reaction to seeing a life-size cast of Lucy is that people note her small size. Every year, I visit an eighth-grade earth science classes taught at our local middle school to talk about human fossils. Among the fossil casts that I bring along is the femur (upper leg bone) of Lucy. I hold it up next to a modern human femur and the students all see very quickly the very small size of Lucy. I then tell them that there are enough of Lucy's remains to get a good idea of her height, which was a bit over 1 meter (3.5 feet) tall.

I ask the students: "Why do you think Lucy was so small?" The answer from some students is that Lucy might have been a child. This is a very reasonable interpretation of her small size because she was about the size one would expect of a five-year-old girl today. Although it turns out that this hypothesis is wrong, it is a good scientific hypothesis because it can be tested. Paleoanthropologists, like forensic scientists, have a number of ways of determining an individual's age at death from skeletal and dental remains. Size is a tricky way to do this when you compare different species. Although 1 meter (3.5 feet) is typical of a young modern human girl, perhaps Lucy belonged to a relatively small species. Instead of relying on size, we look at different measures of maturation throughout the body. For example, when you grow, your long bones (arms and legs) show a progression over time where different bones complete their growth at different ages. In a modern human, for example, growth of the femur is complete by the late teens or early adult years on average. Another example is the bones of the wrist and hands, which complete fusing at different ages, allowing another estimate of an individual's skeletal age. Note that skeletal age may not be the same as chronological age, as some people mature earlier or later than average. Nonetheless, such methods can be used to classify someone's stage of life developmentally.

For Lucy, the best measure we have of her age is her lower jaw. As mammals, humans have two sets of teeth during their lifetime, the deciduous (baby) and permanent (adult) teeth. As you grow, you lose baby teeth and gain adult teeth. The permanent teeth erupt at different ages. The first permanent teeth that typically come in are the first molars on both upper and lower jaws and the central incisors of the lower jaw, all of which on average erupt at about six to seven years of age. Other teeth erupt at different times during childhood and adolescence. For example, a 12-year-old human will typically have all of their adult teeth except for the third molars (wisdom teeth), which erupt between 17 and 21 years of age. As with age estimates from bones, these estimates are of biological age and not chronological age. In addition, the chronological estimates that I have presented are for *modern* humans. Early hominins had dental development more similar to apes, which mature more quickly. Still, we can at least find out whether Lucy was a child, a teenager, or an adult.

Lucy's third molars had fully erupted, which means that regardless of her actual chronological age, she was already an adult. Based on this evidence and the degree of dental wear on the molar teeth, we conclude that Lucy was a young adult at the time of her death. She was not a child and her small size shows that she was a small adult.

However, is this small size typical for her entire species or was she just a small adult on one end of a range of size variation? When considering her size, we also have to consider whether her small size may be due partly because she was female. Many primate species show a great deal of sexual dimorphism, which is the difference in size between adult males and adult females in both body size and size of the canine teeth. In some primates, such as gorillas, there is a great deal of sexual dimorphism. In mountain gorillas, the adult males weigh about 159 kilograms (351 pounds) on average, much larger than the females who average about 98 kilograms (216 pounds). The males also have much larger canine teeth. Other apes show less sexual dimorphism; adult male chimpanzees average close to 52 kilograms (115 pounds) compared with adult females who average about 40 kilograms (88 pounds).[46] The gibbon, a lesser ape from Asia, shows virtually no difference in body or canine size between adult males and females. Present-day humans have some sexual dimorphism, with males being about 15 percent larger than females, but this is far less than seen in gorillas or orangutans.[47]

Given sexual dimorphism, we would expect Lucy to be on the smaller end of the range of variation because she was a female, which is why we call her Lucy, as opposed to Lucius. When possible, sex is determined from fossils using forensic methods, most often applied to the skull or the pelvis. For skulls, males tend to have larger brow ridges and mastoid processes (the bump behind your ear). For pelvic bones, a number of traits can be used to distinguish between males and females. Females having a greater subpubic angle, the angle formed by right and left hipbones at the front of the pelvis; females have wider hips to accommodate childbirth, which produces a much wider subpubic angle. Another trait used to identify sex is the width of the sciatic notch, the notch on the side of the hipbones; here, females have a wider notch. When Lucy's pelvic bones are examined, they are clearly female. Further comparison with other fossils shows that *Au. afarensis* was a small-bodied species to begin with, and Lucy a small female within that species. In fact, Lucy is probably the smallest *Au. afarensis* adult found to date.[48]

How sexually dimorphic was *Au. afarensis*? The size variation in matching parts of the postcranial anatomy suggests that dimorphism was considerable, more than seen in chimps and approaching the level seen in gorillas.[49] However, the sample sizes are still pretty small, making estimates of the level difficult, and some analyses have indicated that the level of dimorphism might be closer to that found in humans.[50] The exact level of sexual dimorphism in *Au. afarensis* is important because of a correlation between body size dimorphism and mating system in primates; monogamous species tend to have little sexual dimorphism and polygamous species tend to have higher levels of sexual dimorphism, presumably reflecting competition between males for females leading to selection for larger males. If dimorphism is high in a fossil species, the inference is that they had a social structure marked by male competition and polygamy, whereas low levels of dimorphism imply monogamy. Further data and analysis will be needed to provide consensus regarding dimorphism in *Australopithecus afarensis*.

Myth 17 *Australopithecus* was a killer ape

Status: At one time, the association of broken animal bones near fossils of Australopithecus *was interpreted as evidence of early tool use, where the animal bones and teeth were used as weapons for hunting other species and for our ancestors killing each other. We now know that the breaks and wear on the animal bones were due to natural causes and not hunting. In fact, the available evidence suggests that* Australopithecus *was often the hunted, and not the hunter.*

One of the most memorable images of early human evolution is found in the opening scenes of the 1968 science fiction film, *2001: A Space Odyssey*, directed and produced by Stanley Kubrick. Set at the "dawn of man," the opening shows an early hominin discovering that a broken bone could be used as a weapon, a very dramatic scene that has the beginning of the powerful composition "Also sprach Zarathustra" playing in the background. The bone weapon is then used to kill animals and later to kill other hominins. At the end of the scene, the triumphant hominin throws his bone weapon into the air and the movie then shifts focus to a space satellite of similar shape, leaving the audience to make a connection between the aggressive nature of our ancestors and the future of our species. (Once a killer, always a killer?)

The idea that violent aggression is rooted in our evolutionary past is a recurring theme in both popular and scientific accounts of human evolution. Another recurring theme that takes a different tack is the idea that our ancestors were peaceful, benevolent, and in harmony with nature. Both of these ideas lead to consideration of a broader question—to what extent can we see evidence of contemporary human behavior in our earliest ancestors? Thus far, we have looked at some of the fossil record for early hominin evolution, such as *Ardipithecus* and *Australopithecus*, forms that were bipedal but still had small brains. What can we tell about their behavior and to what extent did it foreshadow behavior in later hominins, including ourselves?

During the 1960s, the idea developed that *Australopithecus* was a "killer ape" that not only hunted other creatures, but also murdered each other. This view of an ancient root for human violence resonated with some and saw its way into popular culture as well, including the above-mentioned scenes from *2001: A Space Odyssey* as well as the popular book *African Genesis* by Robert Ardrey. In the opening chapter, Ardrey states that our ancestry is "rooted in the animal world" and that among the many behaviors we have inherited, the "most significant of all our gifts, as things turned out, was the legacy bequeathed us by those killer apes, our immediate forebears."[51] In his book, Ardrey specifically describes the evidence for such aggressive behavior in *Australopithecus africanus*, who is seen as a carnivorous hunter that would also murder its own kind. The development of tools and weapons are seen as part of an evolutionary adaptation of *Australopithecus* to feed itself, making humanity an aggressive and competitive group starting millions of years ago.

What is the actual evidence that led to this idea? The roots of the killer ape model lay in large part with discoveries made by Raymond Dart, whom you will remember was the discoverer of *Australopithecus africanus*. In the 1940s, Dart was studying hominin and other animal bones from the Makapansgat site in South Africa, which is not far from other sites where *Australopithecus africanus* had been found. The hominin remains, which we now know date to about 3 million years ago, were considered by Dart to be a species closely related to *Australopithecus africanus*. Because he also found what appeared to be burned animal bones, he concluded that this form of *Australopithecus* had used fire, and subsequently placed the Makapansgat specimens into a new species—"*Australopithecus prometheus*," the southern ape that used fire. Later analysis showed that the blackening of the animal bones was due to manganese staining and not fire, and no further evidence has been found associating any form of *Australopithecus* with the use of fire. Since then, researchers have placed the Makapansgat fossils in the species *Australopithecus africanus* along with the fossils from nearby Taung and Sterkfontein.

What was particularly interesting to Dart were the accumulations of animal bones at the site, including many remains of antelopes and baboons. Dart found that certain antelope bones, particularly humeri (the plural of humerus, which is the upper arm) and mandibles (lower jaws), appeared more often than other types of bones. He reasoned that these bones were being used as tools by *Australopithecus* as part of what he called the osteodontokeratic culture. The term "osteodontokeratic" (which is one of my favorite phrases from paleoanthropology) is quite descriptive because it combined the Greek roots for bone ("osteo"), teeth ("odonto"), and horn ("cerat"), which gives the "bone-tooth-horn" culture. Dart proposed that *Australopithecus* used animal bones, teeth, and horns as tools and weapons. For example, limb bones could have been used as clubs, digging tools, and knives. Shoulder blades could be used as axes, and jaws could be used as knives, saws, and scrapers.[52] Dart also noted that many of these animal bones were broken and scratched, which he thought indicated their use and abrasion.

Dart also noted that the baboon skulls found at Makapansgat looked like they had been killed by a blow to the head, confirming his view that *Australopithecus* was a hunter. Further, Dart saw evidence of violent behavior between our ancestors. Some of the skull fragments of *Australopithecus* showed depressions and fractures that he felt were caused by animal limbs being used as clubs in violent encounters. A number of the lower jaws of *Australopithecus* had their front teeth missing, presumably knocked out in an attack by another *Australopithecus*. Dart's overall assessment was that *Australopithecus* was not only a hunter, but also a murderer and likely a cannibal as well.[53]

Given these interpretations, it is easy to see how the idea of a killer ape in human evolution could take hold. When we see evidence of violent, aggressive behavior in history and in the world today, and then see those same behaviors in our distant past, the conclusion to connect our violent ways to those of our ancestors in an evolutionary framework becomes very tempting. Of course, the suggestion that our violent ways are rooted in ancient evolutionary adaptations opens up questions that are more complex. Are such behaviors innate? Does biology equal destiny? Does an evolutionary origin provide a justification for behaviors that we would otherwise find reprehensible? Are we innately good or bad, and what does this question even mean? Note that such debate also applies to any models that propose the opposite, arguing for an ancient heritage of peaceful, cooperative behavior.

Although Dart made a compelling case for a killer ape, his ideas must ultimately be looked at as hypotheses to be tested. They have been tested and the bottom line is that there is no evidence that *Australopithecus* was a killer ape in the sense that Ardrey and others considered it. Dart made a number of mistakes in interpretation, in large part because little was known about taphonomy, the study of what happens to organisms after they die. What impact does the process of decay and fossilization have on the remains? How can we separate natural from human-made causes for patterns of scratches and fractures? What impact do scavengers have on the distribution and state of fossil remains? These and many other questions have since been investigated through taphonomic analysis.

For example, let us consider Dart's findings that there was an abundance of antelope forelimbs and mandibles. We have to consider that when animals die, we are unlikely to find all parts of their bodies later in time. Smaller bones, such as fingers and toes, will more likely wash away if there has been flooding. Small and fragile bones are also more likely to disintegrate over time. As a rule, we tend to find the harder bones, such as limb bones and jaws, because they are tough and can withstand normal wear and tear, not because they were selected and saved as tools. Further, the bones at Makapansgat have been studied extensively and none of the patterns of breakage or wear is consistent with what we should see if they were indeed being used as tools.[54]

Later research by C.K. Brain on the cave at Makapansgat gives a rather different interpretation from that of Dart. Brain showed that the bones in these caves accumulated through natural processes and had nothing to do with activity by *Australopithecus*. Some bones were likely to have washed in by water and others were brought in by other predators. Bones were also brought in by scavengers, such as hyenas and porcupines, who leave characteristic teeth marks on the bones. Many bones were broken by rocks falling on them.[55] In short, there is nothing at the Makapansgat site that shows deliberate action by *Australopithecus*. From what taphonomy at this site and others tells us is that *Australopithecus* was not a hunter and did not use animal bones, teeth, or horn as tools. The osteodontokeratic culture is not a culture at all, but a collection of animal bones that reflect natural processes.

Brain's work at Makapansgat and elsewhere actually shifted the focus on *Australopithecus* from being a predator to being prey—the hunted rather than the hunter. Evidence to support this view has been found at

South African cave sites. Sometimes early hominins were killed by carnivores and their bones wound up in cave deposits. The caves have vertical entrances formed by water, and there are often trees growing nearby these openings. Leopards will often drag their prey into the trees to eat them, and the leftover bones can drop down the opening into the cave, causing an accumulation of bones over time. In fact, a cranial fragment of an early hominin was found at the Swartkans site with two puncture marks corresponding to the canine teeth of a leopard, which use their canine teeth to drag their prey.[56] Further evidence of animal predation comes from a close examination of the Taung child, the original specimen of *Australopithecus africanus* studied by Dart. Today, monkeys are on occasion hunted by raptors (predatory birds, such as eagles). When this happens, the raptor's talons leave a number of characteristic scratches on parts of the skull and damage to the eye orbits. There are also characteristic scratches that are found when raptors are eating their prey, which are found as well on the Taung specimen. The Taung child has the same type of damage, suggesting that it may have been hunted and eaten by a raptor.[57]

Although the Makapansgat animal bones were not tools, there is some suggestive evidence that some forms of *Australopithecus* might have used stone tools. The species *Australopithecus garhi* has been found at the Bouri site in Ethiopia dating to 2.5 million years ago. Nearby the hominin remains were the remains of a number of mammals, including some that indicated their flesh had been scavenged using stone tools. When a stone flake is used to remove flesh, it leaves characteristic cut marks that can be identified using a scanning electron microscope. Several of these marks have been found on a bovid jaw, consistent with the removal of the tongue. A bovid leg bone shows impact fractures that are characteristic of someone using a stone core to break open bones for their marrow.[58] Although no stone tools have been found at this site, they are known from this time period elsewhere in East Africa. Given that *Au. garhi* is the only known hominin nearby the butchered animal bones suggests (though does not prove) they were tool users. The removal of marrow and tongue is more consistent with scavenging activity, rather than hunting. More recently, stone tools have been found dating back 3.3 million years,[59] and although we do not know who made the tools, it is noteworthy that one or more species of *Australopithecus* lived at this time period. Although it now seems possible that *Australopithecus* may have made and used stone tools, and perhaps was involved in some scavenging, it is still a far cry from the earlier view of *Australopithecus* as an active and proficient hunter, let alone a killer ape.

Although the killer ape hypothesis has been rejected, and it seems that *Australopithecus* was more likely prey rather than a predator, this does not necessarily mean that *Australopithecus* never hunted or was never violent. The absence of wide-scale aggression does not imply completely peaceful behavior. Field studies of apes show that there is a wide range of variation in individual and group behavior, making any hard and fast generalization difficult. Chimpanzees, for example, were initially described as relatively peaceful but we now know that they will kill each other on occasion. They also hunt a variety of animals, including other primates, and have been observed to use sticks as simple spears for hunting smaller mammals. Based on brain size in relation to body size, we suspect that *Australopithecus* was at least as creative and capable as a modern-day chimpanzee, and it would not be a stretch to imagine *Australopithecus* using sticks (or even bones) on occasion as simple tools, or even weapons. As noted above, there is some suggestive evidence for the beginning of stone tool use. However, these observations do not elevate these behaviors to the central driving force as envisioned by those who had proposed *Australopithecus* as a capable and active hunter as well as a frequent killer and cannibal. Instead, we can consider that occasional hunting, and even intra-species violence, existed as part of a wide spectrum of likely behaviors. It is one thing to argue that the *potential* for aggressive and peaceful behaviors exists, and another thing to base the entire course of human evolution on the further development of a killer ape. The osteodontokeratic culture was not a tool tradition, but rather a collection of animal bones, and *Australopithecus* was not a full-time hunter, nor a constant murderer.

Myth #18 Human evolution can be described as a "ladder"

Status: Human evolution is often thought of as a "ladder" consisting of a number of sequential stages leading from past ancestors to modern humans in a straight line. This view proposes that only one hominin species existed at any point in our evolutionary history. One species evolved into the next stage, which evolved into the next stage, and so on. In reality, we now have sufficient fossil evidence to show that there has often been considerable diversity, and more than one hominin species existed at a number of times during the past. Not every hominin was a direct ancestor. An example is the robust hominins classified into the genus Paranthropus— *they belonged to a now-extinct side branch of hominin evolution that lived at the same time as the branch ancestral to modern humans.*

The late paleontologist Stephen Jay Gould has written about the use of "ladders" and "bushes" as metaphors for evolutionary change.[60] A ladder is an object that has a simple linear shape consisting of two straight sides connected by rungs. In evolutionary terms, a ladder conveys the image of a transition from one species to the next in a straight line over time, or what we call anagenesis (see Myth 8). Each species along the family tree is a rung on the ladder. On the other hand, a "bush" is something that has many branches splitting off from a trunk or from other branches, which in evolutionary terms corresponds to cladogenesis. The key difference between ladders and bushes in human evolution is whether one sees all fossil hominins being arranged along a single line from past to present, or whether more than one hominin species ever existed at any single point in time.

When we think of human evolution, we often take the present day as a model for understanding the past. We are the only hominin species alive today, and we might think of possible earlier species as bridging the gap over time between ape and human in a linear sequence. When we combine this tendency with a common bias to see ourselves as the pinnacle of evolutionary progress, a ladder of evolutionary change is an appealing model, because it describes what many see as a steady and linear change leading inexorably to our current state. The ladder metaphor worked well in early days when there were a limited number of hominin species. By the mid-twentieth century, there were only a few players on the field, such as *Australopithecus africanus* and *Homo erectus*, a species with a larger brain than *Australopithecus* but smaller than in modern humans. If we considered only these two species and ourselves, we could easily construct an evolutionary sequence of *Australopithecus africanus* evolving into *Homo erectus*, and then evolving into *Homo sapiens*. This linear sequence corresponded to the observed increase in brain size over time and a limited fossil record. It was a nice neat picture, but one that fell apart when additional species were discovered.

When hominin fossils are discovered, paleoanthropologists must evaluate their similarity and differences with known species. Is it possible to fit new fossils into an already known species, or does a new species need to be named? These assessments are difficult because we cannot determine directly if groups were capable of producing fertile offspring or not. Instead, we look at species differences in terms of anatomical comparisons with reference to the range of variation we typically see in living primate species. If we place two different specimens in the same species, is their difference what we would expect within a single species?

Here, we have to consider multiple sources of variation, including sex differences, age, and the normal kind of variation we see among individuals.

The decision whether to place specimens in an existing species or name a new species also depends on philosophical differences in how common one considers speciation, and under what conditions, as well as starting assumptions about the nature of speciation. Some argue that there is a great deal of variation within living species and we should expect the same levels of variation in the past, and therefore we should start with the assumption that there are few species, each highly variable. Others argue that speciation is very common, and note it is sometimes hard to tell different species from anatomical differences, thus leading us to underestimate the number of species that have lived in the past; as such, we should start with the assumption that new discoveries might be a new species. These two camps are often referred to as "lumpers" and "splitters." A lumper sees a great deal of variation within species and is hesitant to designate a new species, preferring where possible to lump specimens in existing species. A splitter takes the opposite view, seeing speciation as more common and anatomical differences typically reflecting the existence of multiple species.

This may sound chaotic, but in practice, both sides are discussed and considered in different analyses, and there is often a consensus that can be reached over time. In some cases, the differences between two specimens (or samples of specimens) is so great that no one would place them in the same species, whereas in other cases the differences are so minor that there are no question that the specimens belong to the same species. Debate tends to focus on those specimens that are somewhat different, but not extremely different.

In the case of human evolution, there is both consensus on some points and debate over others. As a result, lists of species can vary from one person to the next. Figure 2.5 presents a general middle-of-the-road approach with a list of different hominin species that have been identified for the fossil record over the last 6+ million years. Not everyone will agree with this list. Some would lump some of the listed species into other species, such as placing the species *Paranthropus aethiopicus* (more on this one in a moment) into the species *Paranthropus boisei*. Others would split some of the listed species into additional species; for example, many advocate splitting the species *Homo erectus* into two (or three) different species. Thus, if you were to go to different anthropologists and ask for a similar list you may not get the same exact number. Some would have a

Age range (millions of years ago)		
0–1	*Homo sapiens* (0–200 ka) *Homo floresiensis* (60–100 ka) *Homo neanderthalensis* (28–130 ka)	*Homo heidelbergensis* (200–800 ka) *Homo erectus* (345 ka–1.9 Ma)
1–2	*Homo erectus* (345 ka–1.9 Ma) *Homo habilis* (1.4–1.9 Ma) *Homo rudolfensis* (1.9 Ma)	*Paranthropus robustus* (1.4–2 Ma) *Paranthropus boisei* (1.4–2.4 Ma)
2–3	*Australopithecus sediba* (2 Ma) *Australopithecus garhi* (2.5 Ma) *Australopithecus africanus* (2.3–3.3 Ma)	*Paranthropus robustus* (1.4–2 Ma) *Paranthropus boisei* (1.4–2.4 Ma) *Paranthropus aethiopicus* (2.5 Ma)
3–4	*Australopithecus afarensis* (3–3.7 Ma) *Kenyanthropus platyops* (3.5 Ma) *Australopithecus anamensis* (3.9–4.2 Ma)	*Australopithecus deyiremeda* (3.3–3.5 Ma) *Australopithecus bahrelghazali* (3.5 Ma)
4–5	*Australopithecus anamensis* (3.9–4.2 Ma) *Ardipithecus ramidus* (4.4 Ma)	
5–6+	*Ardipithecus kadabba* (5.2–5.8 Ma) *Orrorin tugenensis* (6 Ma) *Sahelanthropus tchadensis* (6+ Ma)	

Figure 2.5 Summary of proposed hominin species in increments of 1 million years. Note that some species are listed in more than one time period. Note that not all anthropologists accept the validity of some of these listed species (such as *Homo heidelbergensis* and *Homo neanderthalensis*, preferring to view them as variants of *Homo sapiens*). Dates in parentheses indicate the actual range of dates for each species. Most dates are from Relethford (2013b) with the addition of dates for *Australopithecus bahrelghazali* from Cartmill and Smith (2009) and the species *Australopithecus deyiremeda* from Haile-Selassie et al. (2015). The most recent date for *Homo erectus* was taken as the midpoint of the possible range indicated by Indriati et al. (2011). The most recent date for *Homo floresiensis* is from Sutikna et al. (2016).

few less species, and some would have a few more. There would even be some difference in the exact name used, where some would place a given species in a different genus. However, you would also see a great deal of consensus. Regardless of whether one leans toward being a lumper or a splitter, the one thing that everyone agrees upon is that it is not possible to place all of the species discovered into a single evolutionary line as rungs on a ladder. Even if one is an ardent lumper and considers some of the species listed to be variants of a single species, there is still too much variation to be able to reduce the entire fossil record to a single straight line of ancestry and descent.

One of the most diverse periods in hominin evolution was between 2 and 3 million years ago in Africa. (All hominin evolution took place in

Africa before 2 million years ago.) As shown in Figure 2.5, six different species are typically recognized in this time period. Three of these are placed in the genus *Australopithecus*: *Au. africanus*, *Au. garhi*, and *Au. sediba*. All three are small-brained bipeds and differ primarily in terms of their teeth. Each of the three has been suggested to be the direct ancestor of the genus *Homo* and discussion is ongoing about their evolutionary relationships to *Homo* and to each other. Nonetheless, there is consensus that there are differences between these three forms to consider them as separate species. Even if this proves not to be the case and some can be lumped, there are also three other species in Africa during this time that are quite different. These three are listed in Figure 2.5 as belonging to the genus *Paranthropus*, which translates as "beside human," because they are a line that lived at the same time as our ancestors in the genus *Australopithecus*. Although some refer the *Paranthropus* species to the genus *Australopithecus*, everyone agrees that they are different from *Australopithecus africanus* and related species. Here I use *Paranthropus* as a label to help keep the different names straight. The specimens assigned to *Paranthropus* are also often called by the descriptive term "robusts," referring not to body size but to their large back teeth and jaws. Three species are typically recognized based on dental size and certain cranial traits, but here I will discuss them as a general group, focusing on their robust back teeth and jaws.

Paranthropus was a small-brained biped. Like *Australopithecus*, these forms were *megadontic*, meaning they had big back teeth. However, the back teeth (premolars and molars) of *Paranthropus* tend to be enormous with a large amount of surface area for chewing. Some specimens have molar teeth four times the size as those of modern humans. These massive back teeth were placed in large and powerful jaws that were anchored securely onto their skulls. At this point, we need to consider the relationship between jaw muscles and craniofacial anatomy. The upwards force of chewing is accomplished by two set of muscles that shut the jaw when contracting. One set of muscles connects from the lower jaw to the cheekbones. In individuals with very large jaws, such as *Paranthropus*, the cheekbones are very large to anchor the chewing muscles. The other set of muscles goes under the cheekbones (which flare out from the side of the skull) and then anchor to the side of the skull. *Paranthropus* had very large jaws and jaw muscles, and we see this in terms of how much the cheekbones flare out, because larger muscles require more space. In addition, many *Paranthropus* skulls have a ridge of bone running down the center of the skull called the sagittal crest that helps anchor the chewing muscles. Overall, the large jaws (and large jaw muscles) of *Paranthropus*

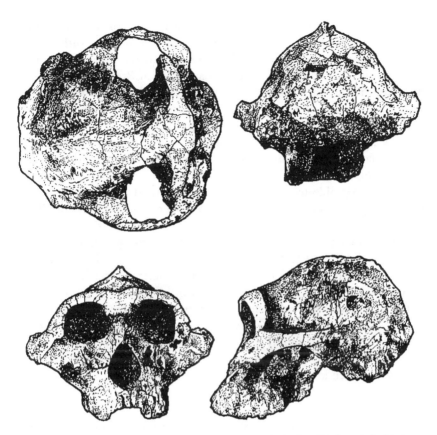

Figure 2.6 Views of *Paranthropus boisei*, specimen KNM-ER-406 from Lake Turkana, Kenya. Clockwise from the upper left: top, back, side, and front views. Note the low skull, wide face with large cheekbones, and the sagittal crest. Source: Cartmill and Smith, 2009. Reproduced with permission of Wiley and Matt Cartmill.

give them a robust facial anatomy, with a large flat face, large flaring cheekbones, and a noticeable sagittal crest in many (Figure 2.6).

What can we make of the robust back teeth and jaws of *Paranthropus*? Ever since the initial discovery of the robusts in the 1930s, the focus has been on the adaptive value of their massive back teeth. As the back teeth are for chewing, we have long considered *Paranthropus* as a hominin that was capable of eating a diet that was hard to chew, such as seeds, nuts, tubers, and harder fruits. Originally, anthropologists envisioned a diet that was distinctly different from *Australopithecus*, which was thought to have been a more generalized omnivore. Isotope analysis of *Paranthropus* teeth and examination of wear patterns on their teeth now show that their diet was more diverse than once thought, but not that different on average from *Australopithecus* who ate softer foods. The current interpretation is that the specialized dental anatomy of *Paranthropus* reflects its adaptation to hard times.[61] The average diet throughout much of the year was the type of food that *Australopithecus* also ate, but they fell back on harder food items when their preferred diet was unavailable. (This is an example of what ecologists refer to as "fallback foods.")

We typically view *Paranthropus* as a descendant of some earlier species of *Australopithecus* that had adapted to changing environmental conditions by being able to exploit other types of food in hard times. As a group, they were successful, as they lasted over a million years, which is respectable longevity for a mammalian species. However, they eventually became extinct about 1.4 million years ago, and have not been seen since then. Why did they become extinct? One intriguing possibility is competition with another type of hominin, as early species of the genus *Homo* lived during the same time period as did *Paranthropus*.

The key point here is that both *Paranthropus* and *Homo* are descendants of species of *Australopithecus*, although we are still working out the specifics of a detailed family tree. As shown in the timeline in Figure 2.5, species of *both* hominin lines lived at the same time. We, of course, are the surviving descendants of the *Homo* line. *Paranthropus* was not a direct relative of ours, but a more distant cousin.

Even if we are taken to very extreme lumping of hominin species, we still would wind up with at least two different lines—*Paranthropus* and *Homo*—showing that a simple "ladder" model is not appropriate as a description of human evolution. The overlap of *Paranthropus* and *Homo* is not the only evidence that hominin evolution is described more appropriately as a "bush" with many branches. It is possible that there were several different lines of the earliest hominins (such as *Sahelanthropus*, *Orrorin*, and *Ardipithecus*). As noted later, in Myth 21, there may have been as many as three different species of early *Homo* in Africa about 2 million years ago. Closer to the present, there may have been several different hominin species within the past several hundred thousand years.

The exact shape of the bush and the names of all the different branches are still being worked out, and will likely get even more complex if more species are discovered. Regardless, we are clear that there have been times when more than one hominin species lived at the same time, and that we cannot describe human evolution as a simple ladder with a set of rungs connecting one stage to the next.

Myth 19 All hominin species have probably been discovered

Status: New hominin species continue to be discovered. Twelve new species of hominin have been proposed since the early 1990s. Even if some of these are variants of other species and should be lumped, there is still evidence of an increase in species diversity. It may be unrealistic to assume that we have already sampled the full diversity of ancient hominins.

There was a time when I could mention at least briefly all of the different species of hominin in my introductory course in biological anthropology. This has not been the case in the past two decades, as the accumulation of data on newly discovered hominin species has increased dramatically. The good news is that we have a more complete picture of our past. The bad news (at least in terms of preparing lectures for my introductory course) is that it is very difficult to summarize this diversity in a handful of lectures. On occasion, I have had students ask me if we are done finding hominin species, to which I reply that this is unlikely to be the case.

When I was in graduate school in the late 1970s, one of my doctoral qualifying exams was in the area of human evolution, specifically debates over the evolutionary relationships of early hominins. As is typical for a doctoral qualifying exam, I had to read and discuss original papers from the scientific literature and to review the broad brush of what we knew about human evolution. At that time, most anthropologists would summarize the hominin fossil record in terms of a limited number of species spanning 3 million years, specifically *Australopithecus africanus*, one or two robust species (*Paranthropus*), and three species in the genus *Homo*—*Homo habilis*, *Homo erectus*, and *Homo sapiens*. Much of the debate at that time dealt with whether the Neandertals, an extinct form of archaic humans, was a subspecies of *Homo sapiens* or should be classified as a separate species. Another debate focused on the validity of the species *Homo habilis*, a form that was intermediate in many ways between *Australopithecus africanus* and *Homo erectus* (and to be discussed in later myths). *H. habilis* had been first proposed in 1964 by the famed paleoanthropologist Louis Leakey and colleagues. Over a decade later, there was still debate over whether a different species designation was actually needed or whether these fossils could be comfortably lumped into other species. The general tone of the time was that proposing a new species required a great deal of justification.

Fourteen years passed before another hominin species was proposed—*Australopithecus afarensis* (and just in time to be included in my doctoral exam!). Again, there was a certain amount of discussion regarding the validity of the species, with some arguing that it might be possible to lump the fossils in an existing species and others proposing that the fossils likely represented two different species. Either way, everyone took note of the fact that the *Au. afarensis* fossils extended both our time depth for human evolution (back to 4 million years ago) as well as our understanding of anatomical diversity in early hominins. Lucy and her kind showed a number of traits that were more primitive than had been found in *Au. africanus*.

Adding *Australopithecus afarensis* to the list did not close the book on what we knew of past species. Twelve new species have been proposed after the early 1990s: *Ardipithecus ramidus* (1994), *Australopithecus anamensis* (1995), *Australopithecus bahrelghazali* (1996), *Australopithecus garhi* (1999), *Kenyanthropus platyops* (2001), *Orrorin tugenensis* (2001), *Sahelanthropus tchadensis* (2002), *Ardipithecus kadabba* (2004), *Homo floresiensis* (2004), *Australopithecus sediba* (2010), *Australopithecus deyiremeda* (2015), and *Homo naledi* (2015). Whew! This long list does not include growing acceptance of other species that had been proposed earlier, such as *Homo heidelbergensis* or *Paranthropus aethiopicus*, or those proposed earlier but not yet accepted by many, such as *Homo ergaster* (a name proposed to separate what others call early African *Homo erectus*).

Why has there been an increase in the number of hominin species? To some extent, the increased number of proposed species might reflect a shift away from a dominant "lumping" philosophy that was apparent decades ago to one more accepting of "splitting." There has also been a shift by many to viewing the evolutionary process as being best described as cladogenesis (bushes) rather than anagenesis (ladders). As these views changed, the atmosphere became more conducive in some ways to proposing and accepting new species.

However, we should not view the increase in species as being caused primarily by shifts in evolutionary biology and philosophy. Regardless of what names are given to these fossils, no one can argue that we have seen a real accumulation in anatomical diversity, regardless of whether we choose to label that diversity with different species names. First, many of these new species date back before 3 million years ago. Before 1978 (the year *Australopithecus afarensis* was announced) we had very few fossils older than about 3 million years ago, and these few fossils were too fragmentary to shed much light on our evolution. Thus, we have doubled the time span of the fossil record for hominin evolution in the past two decades or so. We now have a number of fossil forms, even though still not well known, that extend our knowledge back to 6 million years ago, roughly the predicted time of the ape and human split.

The new fossils have also expanded greatly what we know about early hominin anatomy, showing many examples of different mixes of primitive and derived traits. Although we may argue about how best to classify this variation, it is nonetheless real and, in many cases, of a degree unexpected based on our understanding of human evolution that had existed in the middle of the twentieth century. We are also now seeing unexpected diversity in the later hominin fossil record, such as the surprising discovery of the "Hobbit," a diminutive human-like creature that lived only 60,000 years ago (see Myth 35). New sites have been discovered for many species that further increase our knowledge, such as the finds of early *Homo erectus* from the Dmanisi site in Georgia in western Asia.

Why was all this not known much earlier? Why has so much been discovered only recently? It helps to remember that paleoanthropology is a relatively new science; the first human fossils discovered that were not modern *Homo sapiens* were only found in the middle of the nineteenth century. In addition, paleoanthropology remains a relatively small field even today, with far fewer practitioners than in many other scientific fields. The nature of the research has also changed. In earlier times, paleontological research was most often something done as a sideline to one's primary occupation. Many of the earliest researchers were anatomists by training and would work on the human fossil record in spare moments. Some, such as Raymond Dart, originally relied upon his students and others to bring him interesting specimens to analyze, which is how the Taung child was discovered. Research was also often limited by geography, as many people conducted research near where they lived and did not mount expeditions to faraway places. For example, Dart did his research where he lived and worked, which was in South Africa. There were exceptions of course, but of a limited nature. Those that spent more time in the field, such as Louis and Mary Leakey working in East Africa, were few in number and worked in small groups.

The entire nature of paleoanthropological fieldwork began to change rapidly by the end of the 1960s with the development of long-term studies and international and interdisciplinary teams of researchers. Instead of the lone fieldworker digging through the ground, new teams included experts in a variety of disciplines including geology, anatomy, archaeology, and others. Information was obtained by specialists in geologic dating, taphonomy, stone tool manufacture, pollen analysis, paleobotany, zooarchaeology, and many other fields. New technologies provided more ways of locating and analyzing fossil sites. Computer technology and development of new statistical methods provided better ways of analyzing data. Medical imaging technologies such as CT scanning allow us to look inside fossils. Today, we even supplement the fossil and archaeological records with DNA sequences obtained from ancient DNA analysis. As our tools increase, so does the information we have available.

Still, it would be naïve to assume that we have sampled more than a tiny, tiny fraction of the fossil record. Fossilization is a rare event and many organisms decompose before their bones become fossilized. Some environments are less conducive to the process of fossilization, such as acidic soil. Some fossils are located in places that are not accessible, either for physical or geopolitical reasons. Even in regions where hominin fossils are more abundant, such as the Great Rift Valley of East Africa, the landscape is constantly changing. Sediments are always eroding away, exposing fragile fossils to destruction from wind and water unless someone happens to discover them in time. Fossils that are missed one year might be gone the next. Even though there are more researchers today than in the past, the number is still fairly small and only a fraction of sites can be examined. Although the pace of discovery has accelerated, it is a safe bet that there are many fossils that will never be discovered.

Apart from the myth that we have discovered everything, there is a related myth that we will eventually do so, and will eventually know everything. This is not the case. The past century or so has revealed many aspects of human evolution. We should expect this process to continue. There seems to be a sentiment among some people that given modern technology and scientific advances, we must have at least most of the data on hand and the final answers are close at hand. I have talked to some folks that really expect that we should know the final answer to all questions in human evolution (or science in general) very soon.

I suspect that we often tend to view the present day as the endpoint in a long process of scientific inquiry. We can look back on the history of any science, including paleoanthropology and see hypotheses tested and rejected in past generations, and then conclude incorrectly that we now have it all correct. This view is premature and naïve. It is easy to look back in time and gleefully point out what previous researchers have done wrong, or what they missed, and then marvel at our current present-day knowledge. However, it is necessary to combine this sense of accomplishment with a sense of humility. It is hard to realize that some current ideas will be overturned, favored hypotheses will be rejected, and new data will be available, but that has been the nature of science. There is an unfortunate tendency to treat science as nothing more than a set of accumulated facts, whereas it is actually a process of discovery. As such, it is wiser to consider the present day as a part of this ongoing process. The idea that one's hypotheses may not stand the test of time is disheartening, but that is the name of the game. Science does not offer final facts as much as a way of uncovering truth.

In terms of the hominin fossil record, we have seen an accumulating amount of information. We use these data to test hypotheses and sometimes reject them; for example, we now know that bipedalism did not appear at the same time as larger brains. Old questions are answered, but each answer inevitably brings up new questions. There is no final state of knowledge. I may be wrong but I predict that more hominin species await discovery and naming. (In a sense, my prediction has already been verified, as the species *Australopithecus deyiremeda* and *Homo naledi* were announced during the final editing of this book!)

Myth 20 There are no transitional fossils in human evolution

Status: A frequent critique of evolution is that we have no evidence of transitional forms that bridge the gap between species. This is untrue, particularly in reference to human evolution. The fossil record of human evolution is replete with examples of species that combine primitive and derived traits and provide an evolutionary "link" between ancestral and descendant species.

A perceived lack of transitional fossils, sometimes referred to as "missing links," is a frequent (though incorrect) argument against evolution. This argument typically states that if we suggest that form A evolved into form B, then there ought to be fossil evidence of this transition consisting of creatures that have characteristics of both A and B. In the language of evolutionary biology, we predict that there will be a transitional form that has a mixture of both primitive traits (such as found in A) and derived traits (such as found in B). We may not be able to link all of the specific transitional species within a group to specific ancestors and descendants, but we seek to find fossils representing the general trend from A to B. For example, when we claim that mammals evolved as a group from an earlier group of reptiles, then we should see evidence of transitional forms that have both reptilian and mammalian traits. Likewise, claims that birds evolved from reptiles suggests that we should find evidence of species that combine both reptilian and avian traits. The same is true for other evolutionary transitions, such as from fish to amphibians, or from apes to humans. Keep in mind that when we use terms here such as "fish," "amphibian," "ape," and "human," we are referring to general characteristics of a group, and not exact representations of modern forms. For example, the first fish did not resemble modern fish in all ways (after all, modern fish represent lines of millions of years of fish evolution), but in some basic defining traits.

One of the hallmarks of the scientific process is examining one's ideas and conclusions for potential problems. You need to be able to admit that you might be wrong and identify pieces of the puzzle that are missing or in error thus paving the way for further hypothesis testing. In other words, you have to be critical of your own ideas as well as others. When Charles Darwin proposed that all creatures are descended from others, he marshaled the evidence then available to him. As a good scientist, he pointed out several potential problems with his ideas, one of which was the relative insufficiency of transitional forms. Darwin asked why "if species have descended from other species by insensibly fine gradations, do we not everywhere see innumerable transitional forms?"[62] Darwin suggested several possibilities for the scant evidence of transitional forms, including the imperfect nature of the fossil record.

Since that time, the question of transitional forms has been answered with numerous examples. The fossil record now includes forms that show the transition between many different groups, such as between invertebrates and fish, fish and tetrapods (four-legged vertebrates), reptiles and birds, and many others.[63] A good example is the fossils collectively referred to as mammal-like reptiles, who were a widespread group of mammals that lived during the later parts of the Paleozoic Era (252 to 541 million years ago). Overall, the anatomy of this group was reptilian, but had the different kinds of teeth and jaws of mammals. The reason we use the term "mammal-like" is that this group cannot be placed into the category of either a modern reptile or a modern mammal. Evolution is a continuous process, which makes labeling transitional forms tricky. Nonetheless, the transitional nature of the anatomy is clear. The mammal-like reptiles have long been extinct, but before they died out, they gave rise to the mammals.

What about human evolution? Many argue that if it is true that humans evolved from an early (not modern) ape, then we should see evidence of transitional forms bridging the gaps between very ape-like creatures and modern humans. We not only have evidence of transitional fossils in human evolution, but these examples provide some of the best examples of transitional forms in the fossil record. Consider in broad terms the fossil evidence described in previous myths for four different genera of hominins ("genera" is the plural of "genus"). We have looked at the general anatomy of *Ardipithecus*, *Australopithecus*, *Paranthropus*, and *Homo*. To start, let us consider the evolution of bipedalism. *Ardipithecus* was an early, very primitive biped that showed adaptations for walking upright on the ground but still had an ape-like lower pelvis, long arms, and a divergent big toe. We modern humans have a fully bipedal pelvis, longer legs, and a non-divergent big toe. When we consider *Australopithecus* and the evolution of early *Homo*, we see evidence of a transition. *Australopithecus* had a more fully bipedal pelvis than *Ardipithecus* and had lost the divergent big toe, but still had relatively long arms and short legs compared with modern humans. By the time we get to the species *Homo erectus* (to be talked about more in the next section of the book), we see the development of long legs and evidence of a bipedal striding gait and the possible beginnings of endurance running. By examining these groups in broad detail, we can see the transitional anatomy bridging the gaps between part-time bipeds, early obligate bipeds, and long-distance walkers and runners.

Analysis of canine teeth provides another good example of the transitional nature of human evolution. As described earlier, apes have large projecting canines and the upper canines wear down on their back edge as the tooth is sharpened against a lower premolar tooth. The canines of modern humans are smaller, do not project, are blunt, and wear down from the top of the tooth. The hominin fossil record shows a transition in the teeth of *Ardipithecus* and *Australopithecus* from being more ape-like to becoming human-like. The species *Ardipithecus kadabba*, dating to 5.2 to 5.8 million years ago, has somewhat primitive canine teeth that are more pointed and projecting, with some opposing wear on the cheek side of the lower premolar. However, the wear on the canine teeth suggests that these earliest hominins had begun losing the canine/premolar complex seen in apes.[64] The size and shape of the canine tooth continued to become more human-like in *Ardipithecus ramidus*, and more so in early species of *Australopithecus*, such as *Australopithecus anamensis* and *Australopithecus afarensis*.[65]

Examination of the teeth of *Australopithecus afarensis* provides another excellent example of an evolutionary transition.[66] The canine teeth of *Au. afarensis* are generally smaller than in apes but larger than in humans. The diastema (the gap adjacent to the canines) of *Au. afarensis* is also smaller than in apes but larger in humans. The first of their lower premolar teeth has two cusps, like modern humans, but one cusp tends to be larger than another, more similar to an ape-like condition. Later species of *Australopithecus*, such as *Au. africanus*, have canine teeth that are functionally and anatomically the same as the genus *Homo*. Thus, we see an excellent record of canine reduction in the hominin fossil record between roughly 6 and 2 million years ago.

Another example of an evolutionary transition in human evolution is brain size. As discussed in Myth 12, the earliest hominins still had a small brain size compared with modern humans. For example, the average cranial capacity for *Australopithecus afarensis* is 446 cc (cubic centimeters), much less than an average of about 1350 cc for living humans. The average for *Australopithecus africanus* is 461 cc. We see intermediate values for the earliest species in the genus *Homo* (to be discussed later): *Homo habilis* = 609 cc, *Homo rudolfensis* = 789 cc, and *Homo erectus* = 959 cc. The later species *Homo heidelbergensis* (sometimes referred to as "archaic *H. sapiens*") is larger, averaging 1230 cc, and the prehistoric samples of *Homo sapiens* (before 10,000 years ago) average 1469 cc.[67] (Note that living humans have slightly smaller cranial capacities, consistent with a loss in body mass.) This brief listing does not convey the more complicated pattern in the evolution of brain size, but does illustrate how the gap between the earliest hominins and us can be bridged by a number of intermediate forms.

As we gather more data from the fossil record, we are in a better position to see even more examples of transitional forms, sometimes allowing us to link other species in a sequence of ancestors and descendants. A good example is the species *Paranthropus aethiopicus*, one of three recognized "robust" species (described in Myth 18) that had massive back teeth. The species name was first given in 1968 to a robust lower jaw found in Ethiopia that was somewhat different from the other East African robust species, *Paranthropus boisei* (a very robust species named after Charles Boise, who had provided funds for early fieldwork by the Leakeys). Not much attention was given to the species name "*aethiopicus*" until paleoanthropologist Alan Walker's discovery of a robust skull in 1985 on the western shores of Lake Turkana in Kenya that dated to 2.5 million years ago. This skull was nicknamed the "Black Skull" because of its coloration due to manganese in the ground. Some anthropologists have viewed this specimen as a very early and primitive example of the species *Paranthropus boisei*, whereas others have proposed that it belongs in the same species as the jaw found in 1968.[68] Remember that when specimens are lumped into existing species that have already been named, the older name applies, which is why we call the Black Skull *Paranthropus aethiopicus*, named after Ethiopia, even though the skull was found in Kenya.

Regardless of name, the Black Skull is interesting because it combines both primitive and derived features. It has a protruding face, small brain size, and primitive cranial base, all traits that are similar to those found in *Australopithecus afarensis*. On the other hand, the Black Skull has the large face, large and flaring cheekbones, and sagittal crest found in *Paranthropus boisei*, making it a transitional form that links early, more primitive hominins such as *Au. afarensis* with the later specialized robust forms, *P. boisei* and *P. robustus*.

Many paleoanthropologists view *Australopithecus afarensis* as the ancestor of later hominins, including both *Paranthropus* and *Homo*. If *P. aethiopicus* represents the transitional species connecting *Au. afarensis* and *Paranthropus*, then what links *Au. afarensis* with *Homo*? We are looking for a species in Africa that would bridge the gap between *Au. afarensis* at about 3 million years ago with the first members of the genus *Homo*, at least 2 million years ago. There are actually several possibilities. The species *Australopithecus africanus* has long been suggested to be this link, as it combines some slight cranial expansion and reduction in the canine size. Another, less known species, *Australopithecus garhi* (whose species name means "surprise") also fits the bill in many ways anatomically and is in the "right place, at the right time."[69] *Au. garhi* dates to 2.5 million years ago in East Africa, the same place that we find *Au. afarensis* and early *Homo*, whereas *Au. africanus* is older and from South Africa. More recently, the species *Australopithecus sediba* (meaning "fountain" or "wellspring") has been proposed as a link between earlier *Australopithecus* and *Homo*. This species is known so far from South Africa dating to 2 million years ago, and combines primitive features such as a small brain and long arms with some derived features of the teeth and skull.[70] It has been proposed that *Au. sediba* might be a link between *Au. africanus* and *Homo*, though this is so far not confirmed.

The point here is that there are several species of later *Australopithecus* that lived between 2 and 3 million years ago that could be the link between earlier *Australopithecus* and *Homo*, or at least be related to the actual link. Although we would like to be able to specify a family tree in complete detail with all branches (both mainline and extinct) labeled, the truth of the matter is that such specificity may not be possible. What we can do, however, is to examine overall stages in hominin evolution and see the transitional nature of the evolutionary trends. We have sufficient detail at present to see several transitional stages. The first hominins, such as *Ardipithecus*, represent a stage where our ancestors had small ape-sized brains and had begun to adapt to bipedalism while seeing a reduction in canine teeth. These trends continue in the earlier species of

Australopithecus, such as *Au. afarensis*, and even further with the later species of *Australopithecus* reviewed above. By this point, our ancestors had adapted further to bipedalism and may have been totally committed to it as a way of life. The canine teeth had lost most of their ape-like nature and there was some small amount of brain growth. The next stage is the origin of the genus *Homo*, characterized by an increase in brain size, further reduction of the face and teeth, and an increasing commitment to a stone tool technology. This stage will be covered in more detail in the myths in the next section. For now, we can see the general stages of human evolution as a compelling demonstration of the nature of transitional evolutionary forms.

Notes

1. Falk (2000).
2. Johanson and Edey (1981) provide a nice, clearly illustrated, comparison of ape and human jaws and teeth.
3. For example, see Pilbeam (1972).
4. For a more detailed history of the rise and fall of "*Ramapithecus*" as a human ancestor, see Lewin (1997) and Tattersall (2009).
5. Good sources of information on *Proconsul* are Walker and Teaford (1989) and Walker and Shipman (2005).
6. The history of molecular anthropology is described by Lewin (1997) and Tattersall (2009).
7. Sarich and Wilson (1966, 1967a, 1967b).
8. Sarich and Wilson (1966: 154).
9. See, for example, Scally *et al.* (2012), Disotell (2013), Jobling *et al.* (2014).
10. Langergraber *et al.* (2012).
11. See the news story by Gibbons (2012).
12. Daegling (2004), Cartmill (2008), Loxton and Prothero (2013).
13. Meldrum (2006).
14. Cartmill (2008: 118).
15. Sykes *et al.* (2014).
16. Cartmill (2008: 117).
17. Ciochon *et al.* (1990).
18. Kennedy (2000).
19. Ciochon *et al.* (1990).
20. Meldrum (2006).
21. Krantz (1986).
22. Daegling (2004).
23. Cartmill (2008).
24. Zolliker *et al.* (2005).

25 An excellent review of the discoveries of *Sahelanthropus tchadensis*, *Orrorin tugenensis*, and *Ardipithecus kadabba* is given by Gibbons (2006).
26 Harmand *et al.* (2015).
27 See, for example, Conroy *et al.* (2000).
28 See Ruff *et al.* (1997) and references in Myth 22.
29 See Suwa *et al.* (2009) for information on the canine teeth of *Ardipithecus*, Johanson and Edey (1981) on canine size in *Australopithecus afarensis*, and White *et al.* (2006) on changes in canine teeth over time.
30 See Tattersall (2009) and Cartmill and Smith (2009) for general reviews of the Piltdown Man hoax, and Lewin (1997) and Reader (2011) for more detail.
31 See Feder (2014) for a discussion of the suspects in the Piltdown hoax.
32 An excellent overview of the anatomy of "Ardi" and the series of papers in the October 2, 2009, issue of *Science* on the *Ardipithecus* discoveries is provided by science writer Ann Gibbons in the same issue (Gibbons 2009).
33 Lovejoy, Suwa, Simpson, *et al.* (2009).
34 Lovejoy, Suwa, Spurlock, *et al.* (2009).
35 Lovejoy, Suwa, Simpson, *et al.* (2009).
36 Lovejoy, Suwa, Simpson, *et al.* (2009: 100).
37 Lovejoy (1980).
38 Rodman and McHenry (1980), Leonard and Robertson (1995).
39 Wheeler (1991).
40 White, Asfaw, *et al.* (2009).
41 Louchart *et al.* (2009).
42 White, Ambrose, *et al.* (2009).
43 Stanford (2003).
44 Krogman (1951).
45 Johanson and Edey (1981) provide an excellent and easy to read overview of the discovery of Lucy and the analysis of *Australopithecus afarensis*.
46 Falk (2000). The chimpanzee weights are the average of two subspecies.
47 Larsen (2003b).
48 Johanson and Edgar (2006).
49 Hammond and Ward (2013).
50 Reno *et al.* (2010).
51 Ardrey (1961: 11).
52 Dart and Craig (1959: ch. 11).
53 Dart and Craig (1959: ch. 9).
54 Klein (2009).
55 Klein (2009), Reader (2011).
56 Tattersall (2009).
57 Berger and McGraw (2007).
58 De Heinzelin *et al.* (1999).
59 Harmand *et al.* (2015).
60 Gould (1977: ch. 6).

61 Ungar *et al.* (2008).
62 Darwin (1859: 171).
63 A good list of transitional fossils can be found in the articles on transitional fossils in Wikipedia: http://en.wikipedia.org/wiki/Transitional_fossils and http://en.wikipedia.org/wiki/List_of_transitional_fossils
64 Haile-Selassie *et al.* (2004).
65 White *et al.* (2006).
66 Johanson and Edey (1981).
67 Data on cranial capacities of fossil hominins from Schoenemann (2013). See Chapter 3, Note 8 regarding data selection.
68 Meikle and Parker (1994).
69 Asfaw *et al.* (1999: 634).
70 See Berger (2013) for an introduction to a series of papers on *Au. sediba* in the same issue.

3 EVOLUTION OF THE GENUS *HOMO*

Modern humans are classified in the species *Homo sapiens* (as well as the subspecies *Homo sapiens sapiens*). We are not the first species in the genus *Homo*, a group typically defined by an increased brain size, a reduction in facial and dental size, and the reliance on technology (specifically, stone tools for most of our prehistory). Anthropologists recognize a number of different species in the genus *Homo*, with the earliest showing evolutionary connections with the early hominins described in Chapter 2. This section of the book examines the biological and cultural trends in the evolution of the genus *Homo*, including the emergence of modern humans.

Myth 21 Only one species of *Homo* lived 2 million years ago

Status: In many ways, the species Homo habilis *is intermediate anatomically between Australopithecus and later species in the genus Homo. Earlier views on* H. habilis *posited that they were simply a transitional form between* H. habilis *to Homo erectus. A more complete fossil record shows that the actual transition was more complex, and there were two or three species of Homo existing 2 million years ago in Africa. We are still figuring out the evolutionary relationships between these species. Although it is still likely that* H. habilis *was the ancestor of* H. erectus, *the two species coexisted for at least half a million years. The beginnings of the genus* Homo *reveals cladogenesis, not anagenesis.*

---

50 Great Myths of Human Evolution: Understanding Misconceptions about Our Origins, First Edition. John H. Relethford.
© 2017 John Wiley & Sons, Inc. Published 2017 by John Wiley & Sons, Inc.

In the previous myth, I described some of the more general transitions in human evolution. In this broad view, the species *Homo habilis* is intermediate in many ways between the later species of *Australopithecus* and the species *Homo erectus*. (The species name *Homo habilis* translates as "able man" or "handy man" because of its association with stone tools.) The brain size of *H. habilis* is on average larger than any species of *Australopithecus*, but smaller than for *Homo erectus*, who in turn has a smaller brain size than *Homo sapiens* does. We see a reduction in *H. habilis* of the face and teeth from *Australopithecus*, but the retention of more primitive limb proportions.

Overall, *H. habilis* is anatomically transitional between *Australopithecus* and later *Homo*, which suggests that it might actually be the ancestor linking the two stages of human evolution. Note that because evolution is more often a bush than a ladder, this means that species that show an intermediate transitional anatomy may not actually be the *specific* ancestral link in a family tree. There might be several closely related species, one of which could be the ancestor while the others could just be anatomically similar to this ancestor. As an analogy, note that you have close genetic similarity with your aunt, but she is not your ancestor. Instead, you and your aunt share common ancestry—your grandparents.

In the case of the evolution of early *Homo*, the species *Homo habilis* (Figure 3.1) has been considered for some time as the oldest species in the

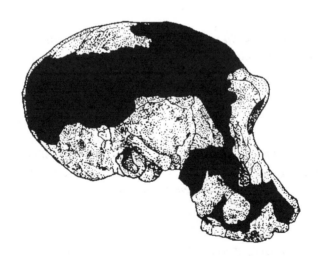

Figure 3.1 Side view of *Homo habilis*, specimen OH 24 from Olduvai Gorge, Tanzania. Source: Cartmill and Smith, 2009. Reproduced with permission of Wiley and Matt Cartmill.

genus *Homo*, dating back to around 2 million years, and perhaps more. This early date, combined with its smaller brain size than other species of *Homo*, has made it a likely candidate for the first member of the genus *Homo*, and by extension the ancestor of more derived species, such as *Homo erectus* and *Homo sapiens*. Following the discovery and naming of *H. habilis* in the mid-1960s, a popular model was to see *H. habilis* as the link between *Australopithecus* and later *Homo* as part of a simple straight-line (anagenesis) evolution—*Australopithecus* evolved into *Homo habilis*, which evolved into *Homo erectus*, which in turn evolved into *Homo sapiens*. This linear sequence is defined by an increase in brain size, a reduction in the size of the face and teeth, and an increasing sophistication in making and using stone tools.

This linear interpretation of the origin of the genus *Homo* was suitable for a number of years until further fossil discoveries showed that the actual evolutionary sequence is more complicated. It turns out that there were two other species that lived in Africa about 2 million years ago that had brain sizes larger than *Australopithecus*. For many years, the default position was to lump any fossils of *Homo* from this time into the species *Homo habilis*. After a while, the growing number of specimens of *H. habilis* from different sites showed more diversity than what might reasonably be expected in a single species. A growing number of paleoanthropologists have argued that more than one species is apparent in the total collection of early *Homo* fossils.[1] Some of these can be assigned to the species *Homo habilis* using the strict and more limiting definition of the species. Other specimens should be assigned to a different species— *Homo rudolfensis*. This species was named after Lake Rudolf in Kenya, an older name for the lake where the fossils were found, which is now known as Lake Turkana.

Homo rudolfensis differs from *H. habilis* in several ways. It has a slightly larger brain case, a skull that is wider in the mid-facial region, a more primitive face, and a larger jaw and back teeth (Figure 3.2). Although *H. rudolfensis* has a slightly larger brain size than *H. habilis*, the more primitive features of the face and teeth have suggested that *H. habilis* was our ancestor and *H. rudolfensis* was an extinct side branch (rather than the reverse). The evolutionary relationship of these two species is complicated by the fact that *H. rudolfensis*, like *H. habilis*, dates back to 1.9 million years ago.

It gets even more interesting. The African forms of *Homo erectus* are now known to date from the same time, roughly 2 million years ago. Although some anthropologists argue that the African forms of *Homo erectus* should be classified as a different species (*Homo ergaster*) than the

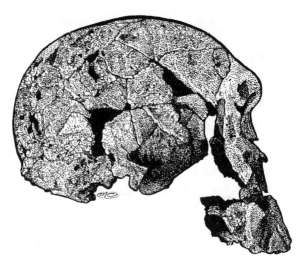

Figure 3.2 Side view of *Homo rudolfensis*, specimen KNM-ER-1470, Lake Turkana, Kenya. Reproduced with permission of Matt Cartmill.

Asian forms, here I will just simply refer to them as "African *Homo erectus*." *H. erectus* has a larger brain size than either *H. habilis* or *H. rudolfensis*. The higher vault of the skull reflects this larger brain size, but the braincase is still smaller than that of a modern human (about 70 percent the size of a modern human), and the skull is more elongated (football-shaped) than our rounder (basketball-shaped) skulls. The face of *H. erectus* is reduced compared with earlier hominins but still protrudes a bit in the lower part of the face and jaws. One of the most noticeable features of *H. erectus* is the massive continuous brow ridges (see Figure 3.3) that form a shelf above the eyes. *H. erectus* also has a more modern postcranial structure with long legs and limb proportions more similar to ours than *H. habilis*, which still has relatively long arms and short legs. Overall, *H. erectus* is more like us than either *H. habilis* or *H. rudolfensis* and we see it as our ancestor from this time. What about the other species of *Homo* from this time? Where do they fit into our family tree?

Constructing a family tree is complicated because we have three species of early *Homo* in Africa dating back to the same time (or at least two, as not everyone accepts *H. rudolfensis* as a separate species). Of these, *Homo erectus* is clearly the one most closely related to us and thus can be considered our ancestor. What about the other two species and their relationship, if any, to *H. erectus*? Obviously, these data do not support a simple model where *Homo habilis* evolves directly into *Homo erectus*

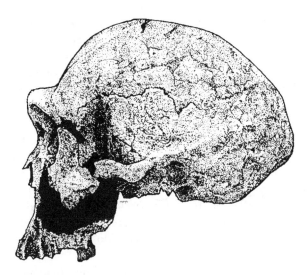

Figure 3.3 Side view of early African *Homo erectus*, specimen KNM-ER-3733, Lake Turkana, Kenya. Source: Cartmill and Smith, 2009. Reproduced with permission of Wiley and Matt Cartmill.

because we see both ancestor and descendant at the same time. However, although we can rule out straight-line evolution (anagenesis), that does not mean that *H. habilis* could not have been the ancestor of *H. erectus* when we consider branching evolution (cladogenesis—refer back to Myth 8). Suppose *H. habilis* existed before 2 million years ago and then part of the species split off to form *H. erectus*. In that case, we might expect to see both the parent species (*H. habilis*) and the daughter species (*H. erectus*) living at the same time, which fits our observations. Regardless of the origin of these species, it is interesting to note that *H. habilis* lived in East Africa until 1.4 million years ago, which means it lived side by side with *H. erectus* for half a million years.[2]

What about *Homo rudolfensis* under this model? One possibility is that it too split from an earlier population of *H. habilis* and was still around 1.9 million years ago before it went extinct. Of course, we could also argue that maybe it was the other way around, with *H. rudolfensis* being the parent species, giving rise to both *H. habilis* and *H. erectus*. Given what we know about these species, this seems less likely, as *H. rudolfensis* appears to be a less suitable candidate for an ancestor of *H. erectus*, leading us to prefer *H. habilis*. To make things even more complicated, consider another possibility where another, thus far unknown, species was the ancestor of all of the three species of *Homo* that we see in East Africa at 1.9 million years ago.

Finally, to add even more to the range of possible scenarios, perhaps one or more of the species of *Homo* evolved independently from earlier species of hominins. Some anthropologists have suggested this possibility due to an interesting similarity of the facial anatomy of *Homo rudolfensis* and the 3.5 million-year-old skull classified as *Kenyanthropus platyops* (translated as "the flat-faced man from Kenya," an appropriate name given the wide and flat mid-facial anatomy of this specimen).[3] The difference between *K. platyops* and *H. rudolfensis* is that the former has a small cranial capacity (estimated between 400 and 450 cc for the one known skull) compared with the latter (an average of 789 cc for two skulls).[4] If *H. rudolfensis* evolved from *K. platyops* while *H. habilis* evolved from a species of *Australopithecus*, then we would be seeing a parallel increase in brain size in two different hominin lines. At present, too little is known about *Kenyanthropus* to do more than note this as an interesting but still untested hypothesis.

Apart from the potential complications from *Kenyanthropus*, we must still deal with determining the possible ancestor of the three *Homo* species at 1.9 million years—was it *Homo habilis*, *Homo rudolfensis*, or a third unknown species of *Homo*? At present, we can state some preference for picking *H. habilis* over *H. rudolfensis* because it is less primitive in some traits. We might also prefer *H. habilis* to an unknown species because it is not a good idea to construct a family tree using data that have not been discovered! True, we might find something, but then again, we might not.

We clearly need more information on the beginnings of the genus *Homo* prior to 2 million years ago. Here we have good news and bad news. The good news is that we *do* have some fossil evidence for earlier *Homo*. The bad news is that this evidence is not conclusive enough because of the fragmentary nature of the specimens (no complete skulls) to assign them definitively to a particular species. One fragment of the temporal bone of the skull dates to 2.4 million years ago and might belong to the genus *Homo*, but not enough data are available to be more specific. In addition, an upper jaw from Ethiopia dates back 2.3 million years ago and can be placed in the genus *Homo*. Although this specimen is a bit more similar to *H. habilis* than to *H. rudolfensis*, we do not have enough information to nail this down. A lower jaw from the site of Malawi between southern and eastern Africa dates between 2.4 and 2.5 million years old, and is similar in some ways to *H. rudolfensis*.[5] Finally, a partial lower jaw dating back 2.8 million years ago has been discovered in Ethiopia that has both primitive features similar to that of *Australopithecus afarensis* but also more derived features of *Homo*.

The discovers have placed it in the genus *Homo*, but there is not enough data to determine the species.[6]

Consequently, we have a tantalizing glimpse of *Homo* prior to 1.9 million years ago, but not enough to sort through the various hypotheses. In my view, the available evidence leans a bit toward the model of *H. habilis* being the ancestor of *H. erectus*, with both parent and daughter species living at the same time, but this hypothesis could easily be rejected given more data. We'll have to wait and see. Another recent discovery that may provide insight is the discovery in 2015 of a possible new species—*Homo naledi*—that has some characteristics of both *H. habilis* and *H. erectus*, as well as some unique features.[7] For the moment, we cannot fit this new species into the broader picture, as it is not yet dated.

Although we cannot sort out the exact nature of the family tree of early *Homo*, we have enough evidence to reject the idea that there was a single species of *Homo* in Africa 2 million years ago. This discussion also provides another good example of the nature of scientific inquiry in paleoanthropology. We have sufficient data to see the big picture and the general trends in human evolution and are now working to resolve the picture at finer detail. As we do so, we are able to reject hypotheses such as the one for a single *Homo* ancestor, which in turn generates further hypotheses that come from the finding that there is more than one species of early *Homo*. Such uncertainty may be unsettling, but that's the name of the game!

Early *Homo* had modern human brain size

Status: Although the genus Homo *is characterized in part by having a larger brain size than early hominins, early* Homo *did not have modern brain size, and the transition to modern human brain size took place over millions of years. We are still exploring possible reasons for the evolution of larger brains.*

The genus *Homo* has long been defined by a large brain size, exceeding that of apes. This increase in brain size, along with a reduction in the size of the face and teeth and the development of stone tool technology, allows us to talk about these ancestors as specifically *human* rather than just hominin. The appearance of larger brains and stone tools marks the beginning of biology and behavior that we would consider human. However, it is important to understand that these early humans (such as *Homo erectus*) are not yet the same, biologically or culturally, as *modern* humans. Thus, the term human can be misleading when used without appropriate qualifiers, such as "early," "archaic," "modern," or some other way to illustrate the range of biological and behavioral variation.

It is important to realize that defining the genus *Homo* as having larger brain sizes does not mean that all humans, early and modern, have the same brain size. The earliest members of the genus *Homo* had larger brains on average than *Australopithecus*, but not as large as later species of *Homo*. In other words, brain size did not increase all at once.

What is the pattern of evolution of human brain size? First off, we must keep in mind that brain size can only tell us so much. When we estimate the cranial capacity of a fossil skull (as described in Myth 12), we are getting an estimate of the *absolute* volume of the interior of the skull. However, absolute brain size varies with body size. Larger species typically have larger brains (compare, for example, an ape with a lemur), and within species the larger individuals typically have larger brain sizes. In many cases, brain size *relative to* body size is a more appropriate measure. Although we can estimate body size from some postcranial remains, we do not always find individuals where we have both the necessary postcranial remains and an estimate of brain size.

Another drawback of absolute brain size is that it does not tell us anything about the structure and organization of the brain, such as the disproportionate growth of some parts of the brain relative to others. We can sometimes get at such information from endocasts. As noted in Myth 13, an endocast is a cast of the interior of the skull, which can tell us something about the surface anatomy of the brain and proportions of different parts of the brain. Some endocasts occur during fossilization (such as the endocast of the Taung child described in Myth 13). Endocasts can also be made using latex to form an impression of the interior of the skull. A more recent development has been the use of medical imaging technologies to create virtual endocasts. Although useful, endocasts are unfortunately not available for all specimens. For some analyses of long-term brain evolution, we have to focus primarily on absolute brain size.

Still, data on absolute brain size are adequate to show easily that modern human brain size did not evolve all at once in the course of human evolution. The estimates of cranial capacity for fossil specimens can be plotted by their geologic age to get an idea of changes over time. This was done in Figure 3.4 for 160 fossil specimens in various species of the genus *Homo* prior to 10,000 years ago.[8] Brain size in *Homo* shows a clear increase over time, more than doubling from the time of early *Homo* to modern *Homo*. The scatter of points in Figure 3.4 shows some curving,

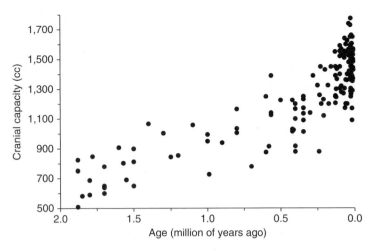

Figure 3.4 Plot of cranial capacity (in cubic centimeters) as a function of geologic age (millions of years) for 160 specimens assigned to species of Homo (*Homo habilis*, *Homo rudolfensis*, *Homo erectus*, *Homo heidelbergensis*, Neandertals, and *Homo sapiens* prior to 10,000 years ago).

and it might be the case that there has been some increase in the brain size starting about 750,000 years ago. Regardless, it is clear that the first members of the genus *Homo* did not have modern human brain size. This is true even if we exclude *Homo habilis* and *Homo rudolfensis*, as suggested by some anthropologists. The evolution of human brain size is an excellent example of substantial evolutionary change in a relatively short (geologically) length of time.

Why did brain size increase in the genus *Homo*? A trend of this type is strong evidence that there has been selection for larger brains. However, what is the nature of this selection? Keep in mind that a focus on overall brain size is crude, and what is likely more important are correlated changes in the structure and organization of the brain. Analysis of fossil endocasts gives us some basic ideas about the evolution of different parts of the brain, including a reduction in the visual cortex, an increase in asymmetry of certain areas, and a likely increase in the prefrontal cortex. In addition, there are human-like patterns in early *Homo* associated with language ability. There is also greater expansion of some parts of the brain than in others, and these changes might be related to spatial ability and motor control.[9] Size does not tell us everything, and we are likely seeing both an increase in overall size as well as brain reorganization over the course of human evolution.

Such changes imply strong selection over time for larger and more complex brains. Why? The behavioral correlates of brain evolution are difficult at best to read from the fossil record, and we must consider evidence of past behavior from the archaeological record as well as inferences based on the study of living primates. (What are apes capable of, and what might that tell us about the capabilities of our ancestors?) It seems rather obvious that larger and more complex brains are related to an increase in cognitive ability and intelligence. However, these are rather broad and vague terms and encompass a wide range of mental abilities that could increase the survival and reproduction of our ancestors in numerous ways.[10] One of the most obvious is generalized problem-solving, which gives the ability to adapt to new circumstances and challenges. A related ability is manufacture and manipulations of technology—toolmaking. In prehistoric times, tools would have been manufactured out of stone, and later in time, bone, as well as some basic uses of simple sticks and stones. We do the same thing today in an enhanced way because our knowledge base accumulates as part of human culture. We do not need to reinvent the wheel each generation. The transmission of culture is helped along by our ability to use language. Some consider the origins of language and its subsequent evolution to be another important (and perhaps most important) correlate of brain evolution.

We can go even farther. Memory would play an important part of survival, such as remembering when and where to forage and scavenge. Spatial ability and an awareness of the environment would also be useful. The evolution of mental abilities was likely not limited to solving problems of the physical environment (such as eating and keeping from being eaten). Social adaptations likely also occurred; some have suggested that large brains represent an adaptation to larger social groups, because of the increased number of social relationships in larger groups. It seems more appropriate to consider the entire range of behaviors and adaptations made possible with a larger, reorganized, and more complex brain, rather than trying to identify a single cause and effect relationship. We may be seeing over time different influences of an array of behaviors including tool use, language, and social complexity, among others.

We have to consider also the potential *disadvantages* of having a larger brain. What are the costs of having a larger brain? For one thing, larger brains consume a great deal of energy. In living humans, the brain makes up about 2 percent of our total body weight, but consumes 20 percent of our energy. An energetically expensive organ might be a liability in situations where food resources are limited, and this disadvantage must be compensated for by the advantage of a larger and more complex brain for acquiring additional food resources, such as afforded by the invention and use of stone tools.

The higher energy requirements of the human brain might explain an interesting fact about the relative size of organs in the human body. When we look across different primate species, we see a close relationship between body size and organ size. For some organs, such as the heart, liver, and kidney, our organs are just about the size you would expect for a primate of our body size. Our brain size, however, is about three times the size we would expect for a primate of our body size. In the language of statistics, we are an outlier with our very large brains (with their correspondingly high energy needs). We are an outlier in another way as well—our gut mass is much less than expected for a primate of our size. Humans have large brains and small guts. Leslie Aiello and Peter Wheeler have proposed that these two facts are related in what they refer to as the expensive-tissue hypothesis, because both the brain and the gut are metabolically expensive. In their model, there was a trade-off between brain size and gut size during human evolution; in order to expend more energy on the brain, less energy was available elsewhere, and smaller gut mass would have lower metabolic needs. They view this trade-off in terms of changes in diet, because you need a diet of high quality foods that are easy to digest if you are going to have smaller guts.[11] In terms of our evolution, this shift may correlate with the addition of animal proteins to the diet through scavenging and processing with stone tools. Aiello and Wheeler suggest that the smaller gut size had been reached by the time of early *Homo erectus*, whose skeletal structure has a more modern thoracic shape that reflects smaller guts (compared with the funnel-shaped thorax of apes and earlier hominins). The important suggestion here is that a significant increase in brain size would not have been possible until a shift in diet allowed the trade-off for energy needs, because we could not afford both large brains and large guts. Another potential disadvantage of larger brains is that they are harder to cool. A well-established principle of mammalian physiology is that large organisms or larger structures (such as the head or limbs) lose heat less rapidly than smaller organisms or structures.

Another problem with larger brains is childbirth. In order for a human to grow up to have a large brain, some of the brain growth has to occur before birth. However, the more the brain grows during the final trimester, the bigger the brain is at birth. This poses a problem in humans because of the changes to the pelvis that allow us to be bipedal. Thus, it is much more difficult for a human mother to give birth to a large-brained infant than is the case for an ape mother giving birth, with a differently shaped pelvis and babies with smaller heads. The potential mortality to both mother and child when giving birth to a large-brained baby is another major disadvantage of having big brains. Humans have evolved a partial solution in that rapid brain growth after birth continues longer in humans than in apes. However, this adaptation poses its own problems because human infants are then more dependent and fragile for a longer amount of time in infancy and childhood, thus requiring their mothers to spend more time and energy on them.

As noted in several other myths, in evolution nothing is free. Larger brains have both advantages and disadvantages, and any increase reflects conditions where the advantages outweigh the disadvantages giving an increase in net survivorship and reproduction. It is clear that there are also constraints on how much the brain can increase, such as available energy. It is likely that the balance between cost and benefit shifted when early hominins began extracting animal products (meat and bone marrow) using stone flakes. This simple adaptation was possible in a primate whose past evolutionary history had provided keen eyes, free hands, good hand–eye coordination, and relatively large brains compared with many mammals. *Homo habilis* had a brain size larger on average than *Australopithecus*, but it is not clear how much of this increase was due to an increase in overall body size. A substantial increase in brain size is seen in *Homo erectus*. Perhaps as *H. habilis* increased the amount of animal products in their diet, this solved in part the energetic needs of larger brains, opening the door for even further increase in the size of the brain, as well as a corresponding reduction in gut size, both of which are seen in *H. erectus*.

In any event, there came a time when our ancestors began to rely increasingly on learned behavior and tool use, which in turn selected for even more problem-solving, increased mental abilities, and larger and/or reorganized brains. In turn, the energy demands of larger brains and the need to care for immature infants set up conditions that favored even more reliance on tools and cultural behaviors.

Myth #23 Only humans are toolmakers and have culture

Status: For a long time we thought that toolmaking and cultural behavior (in the sense of shared and learned behavior) was an exclusively human activity. Studies of chimpanzees and other primates show that nonhuman primates use tools and have culture in the broad sense. Human culture is still qualitatively different because of its reliance on language acquisition, a behavior that apes approach, but do not obtain in full.

Regardless of the anatomical and genetic evidence of our relationship to the apes, we often try to distance ourselves from the rest of the natural world. There has been a long tradition throughout history of seeing ourselves either at the top of the ladder of living things or separate from nature altogether. We consider ourselves special, be it in a philosophical or scientific sense. After all, anthropology is the study of humans, one species out of countless millions of species in the world. I am not even sure I can explain why I wanted to spend my career studying humans (and close relatives, past and present) rather than plants, insects, dogs, or cats (although I must admit that dinosaurs have always come in at a close second in terms of personal interest).

As an anthropologist, I would argue that a single focus on our own species is a normal extension of people's desires to connect in one way or another to their own history, be it the history of our family, community, nation, or species. Two of the natural questions that we then ask are, how are we similar to other creatures, and how are we different? We can explore these questions in terms of both our biology and behavior. We have examined this question in previous myths for biological traits such as bipedalism and large brains. We can also ask the question in terms of behavior. One long-standing answer to the question of behavioral uniqueness is that we are toolmakers. In fact, the phrase "man the toolmaker" was quite common many decades ago, and it summarized what we thought was a unique feature of human evolution.

We use "tools" in a very broad sense and not in the more specialized sense such as hammers, screwdrivers, and other bits of hardware you might buy in a hardware store. Toolmaking in the broad sense means modification of some part of the environment (other than one's self) and using it to solve a problem, such as getting food. To some extent, even entertaining the notion that only humans are toolmakers seems absurd when our general definition can include the facts that birds build nests, beavers build dams, and other examples. However, is this the same thing? Most would argue that human toolmaking is a behavior that is more open-ended than seen in other species, and more centered on problem-solving.

The view that this open-ended, problem-solving type of toolmaking was unique to humans changed dramatically beginning in the 1960s when it was discovered that chimpanzees made and used simple tools in the wild. These observations were first made during the pioneering work of Jane Goodall, who studied chimpanzees in their native habitat, the rain forests of western Africa, and not in the artificial environments of labs or zoos.[12] Goodall was the first to observe termite fishing among chimpanzees. Chimps enjoy eating termites, which are difficult to capture once they have moved down long tunnels into their nests deep inside large mounds of dirt. Goodall observed chimpanzees taking a stick or grass stem, sometimes stripping off extra leaves, and then using these sticks to probe gently into the tunnels. When the stick reaches the termite's nest, the termites view it as an invader and latch onto the stick with their jaws. The chimpanzee will then withdraw the stick very slowly and carefully, and then eat the termites off the end of the stick. This sounds rather simple, but the entire process is something that does take some practice. You have to pick the right type of stick, as one that is too hard or too flexible will not work when trying to follow the tunnel into the nest. Of particular significance is the fact that chimpanzees will sometimes strip leaves off a stick to render it useful, which is, by all definitions, an act of toolmaking as well as an act of tool use.

Termite fishing is a learned behavior. It is not something that is innate, as we do not see it in all chimpanzee groups located near termite mounds. Chimps learn how to termite fish, as shown by the observation that young chimps observe and imitate the termite fishing behavior of their elders. This observational learning provides a way for termite fishing behavior to be passed from one generation to the next.

Since the time of Goodall's initial studies, many other examples of tool use and toolmaking have been seen in chimpanzees (and some in other apes). Chimpanzees will chew leaves partially to create an absorbent mass that they use to soak up water from holes in tree branches. They also use stones to crack open nuts, and sticks to fish for ants. Bone splinters are used to pick marrow out of bones, and leaves are on occasion used as toilet paper. In several cases, chimps have been observed to strip leaves off a stick, sharpen the end with their teeth, and then thrust their spear into a tree hole to kill smaller primates.[13]

Apart from demonstrating their creativity and intelligence, studies of chimpanzee behavior provide ample evidence that tool use and toolmaking, as well as other behaviors, are learned in a social group. Do these learned behaviors mean that chimpanzees have culture? A basic definition of culture is shared and learned behavior. Is this what we see in chimpanzees? To answer this, a landmark study in 1999 combined the observations of chimpanzee behavior over several decades at seven different chimpanzee groups in Africa.[14] The scientists pooled their data to determine which behaviors were common to all chimpanzees, which were shared by some but not all groups, and which were unique. They examined 65 different behaviors that included the above examples of tool use (and many more) as well as other behaviors, such as performing a "rain dance" as a display behavior and holding hands overhead while grooming. The purpose of the study was to find behaviors that were evidence of habitual behaviors found in some but not all groups and whose absence in some groups was not due to environmental factors (for example, you cannot fish for termites if there are no termite mounds).

The researchers found that seven of the 65 behaviors were found often in all seven chimpanzee communities, such as dragging large branches as part of an aggressive display. An additional 16 behaviors were found to be relatively rare in all chimpanzee groups. Three other behaviors were absent in many groups because of environmental factors; for example, although one chimpanzee group uses stripped stems to scoop up algae, the others do not because there is no algae available where they live. The remaining 39 behaviors studied are the ones that are significant for our understanding of culture in chimpanzees because they were found in some groups but absent in others. For example, termite fishing was found in some groups but not others, even though there were termite mounds near some groups that lacked this behavior. Likewise, chimpanzees in some groups clasped their hands together over their heads while grooming, but this was not seen in other groups. Such group-specific behaviors are learned and not innate. Although there have been numerous examples of single behaviors that are socially learned in a number of animal species, only chimpanzee groups show sets of learned behaviors. Each group had its own constellation of learned behaviors, a pattern typical of what we see in human culture.

Does this mean that chimpanzees have culture? If so, then, as with toolmaking, we can no longer use the term to be exclusively human. The answer to this question depends on whether you follow a broad or specific definition of "culture." If we take culture to refer simply to shared and learned behavior, then chimpanzees can indeed be said to possess culture. On the other hand, anthropologists tend to define culture as involving language in the transmission of learning. Whereas both humans and chimpanzees use observation and imitation to learn something new, only humans facilitate this process with language. When we show our children how to tie their shoes or brush their teeth, we use language (spoken and gestural) as well as observation for instruction. Chimpanzee young will watch their elders termite fishing and then imitate them, perfecting their ability through trial and error. If we humans were to show our children how to fish for termites, we would likely combine the physical instruction with a long commentary describing how to select an appropriate stick, how to insert it into the mound, and other instructional aides. Our linguistic ability allows us to transmit knowledge from one generation to the next for both simple and complex behaviors and achievements. Although observational learning and imitation is sufficient for some behaviors, language is necessary to farm, build pyramids and steam engines, and to discuss philosophy.

Is it the case, however, that only humans use language? To answer this, we have to consider the difference between communication (which chimpanzees as well as many animal species have) and language. Language is typically defined as symbolic communication that has a number of particular characteristics, such as being open (allowing you to make up new symbols and ideas), generalized (extending symbols to new situations), and capable of displacement in space and time (communicating about things not present or something in the past or future), among others. Chimpanzees make sounds and gestures that have meaning, but they are more limited. A chimpanzee might hoot to indicate imminent danger from a predator, but humans can indicate far more with language. We can indicate the exact nature of the threat ("There are three lions ahead in the grass"), severity of the threat ("They are approaching us very quickly"), and the response to the threat ("We need to climb the large tree to our left"). We could also talk about past events ("Last week we were able to chase them off with rocks") and the future ("We need a better warning system for lions!"). When the danger has passed, we could use the experience as a teaching lesson or incorporate it into a myth. Although it turns out that nonhuman animal communication systems are more complex and nuanced than we once thought, it is also clear that we are alone in our level of linguistic mastery. Chimpanzees in their native environment do not use language in the sense that humans use language.

However, does this mean they are incapable of learning language? One of the goals of understanding human evolution is to note what differences exist between human and ape, and use this comparison to figure out how our unique traits might have evolved. In the case of language, was it a gradual process of change, building off abilities in apes and evolving from these beginnings through *Australopithecus* and *Homo*, or did humans acquire the ability to use language in a short interval of evolutionary time? In either case, it is useful to know the baseline condition for language acquisition that was likely to exist in a common ancestor based on the level of ability that we see in living apes.

Early attempts to teach a chimpanzee to speak English failed. In retrospect, this failure is not surprising because apes lack the anatomical specializations that allow humans to use a spoken language. However, does that necessarily mean that apes are also mentally incapable of learning a language? Pioneering studies in the 1960s centered on teaching a chimpanzee to learn American Sign Language (ASL), a gestural language that has all the symbolic features of a spoken language, but requires gestures with the fingers and hands, as well as facial expressions, all of which are possible in apes. The first ape ever taught ASL was a young female named Washoe, who learned a large number of signs and showed the ability to generalize the meanings of signs. Washoe was capable of simple two- or three-word sentences in ASL, such as "gimme tickle." Since that time, other symbolic languages have been invented and used to teach apes, some using plastic tiles and some using a large computer keyboard. Language acquisition experiments have been conducted using a number of chimpanzees as well as other species of apes, such as gorillas and bonobos. In general, apes demonstrate basic understanding and can communicate with humans to a limited extent with very short sentence fragments. There is still debate over the extent to which apes actually understand the underlying rules of language, particularly grammar and sentence structure, or are simply learning correct responses to humans.

Regardless, the ape language studies show that apes are more capable than was thought in the middle of the twentieth century.[15] It is also clear that apes have a limit in their language acquisition abilities, so no one will be having intricate discussions with apes regarding the state of the economy, the meaning of life, or why they enjoyed a movie. However, their capabilities, even though limited, show potential that might have also existed in a common ancestor, a potential that was realized and expanded upon during the course of human evolution.

In sum, the behavioral nature of humanity has often been defined in terms of toolmaking, culture, and language. These behaviors have often been viewed as distinctively unique to us. Studies of toolmaking and culture in apes show us that the difference between them and us may be more a matter of degree than kind. However, apes are not humans, and the distinctive nature of humans is apparent when seeing the connection between culture and language that has developed during human evolution.

We can identify species by the stone tools they made

Status: The evolution of the genus Homo *is characterized by an increase in the size of the brain and a reduction in the size of the face and teeth. There has also been evolution of stone tool technology, starting with simple chopping tools that are followed by increasingly more sophisticated stone tools. However, even though later species of* Homo *invented new types of stone tools, these technological changes do not occur at the same* time as the biological shifts. New species typically continue using earlier types of tools. Consequently, we cannot always determine who made stone tools unless we also find their fossil remains.

The story of human evolution is not just the story of our biological changes, but also our cultural changes. Over the past 2 million years, the genus *Homo* shows an increase in brain size, a reduction in the size of the face and teeth, and an increasing reliance on stone tool technology. Stone tools change over time, becoming more sophisticated, more complex in manufacture, and more diverse in application. We also see evidence in the archaeological record for geographic expansion, increased hunting ability, use of fire, and the development of symbolic behavior such as art. There is no doubt that the genus *Homo* has evolved both biologically and culturally, and new species are associated with new types of stone tool technology, which makes sense in terms of the evolution of cognitive ability.

However, the actual pattern of change over time is a bit more complicated than might first be apparent. For example, although new species of *Homo* invented new types of stone tools, they did not do so from the very beginning of their species' existence. New species frequently use the tools of previous species for a while and later develop new technologies. Biological and cultural evolution was not synchronized and there is often a lag between biological and cultural change. This means that when we find stone tools but no fossils we cannot always tell *who* made the tools.

To show this, we start with a very brief review of some of the major events in the evolution of the genus *Homo*, some of which will be explored in more detail in later myths. In Myth 21, we saw that the genus *Homo* arose in Africa and, by about 2 million years ago, there were three species of early *Homo*: *Homo habilis*, *Homo rudolfensis*, and *Homo erectus*. Populations of *H. erectus* dispersed into western Asia (Georgia) by 1.75 million years ago and southeastern Asia by around 1.7 million years ago. *H. erectus* later moved northward into east Asia, where they lived until about 400,000 years ago. The populations in southeast Asia persisted until sometime between 143,000 and 546,000 years ago. (The dating method used here provides a range of dates.)[16] Before *H. erectus* died out, it gave rise to populations characterized by a larger brain, marking a transition between *H. erectus* and *Homo sapiens*. Although some paleoanthropologists refer to this stage as "archaic *Homo sapiens*," there is growing consensus to treat it as a separate species—*Homo heidelbergensis*, named after the city of Heidelberg, Germany where the earliest specimen was discovered. The transitional nature of *H. heidelbergensis* is clear (Figure 3.5); it has a larger brain size than *H. erectus* (but still about

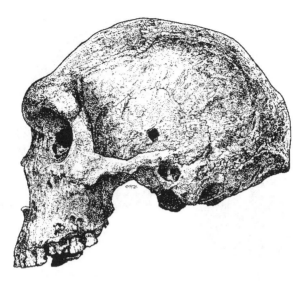

Figure 3.5 Side view of *Homo heidelbergensis*, Broken Hill I specimen, Zambia, southern Africa. Source: Cartmill and Smith, 2009. Reproduced with permission of Wiley and Matt Cartmill.

10 percent less than *H. sapiens* in absolute size), and still has a large face and a football-shaped skull. The brow ridges remain large, although are more separated, unlike the continuous shelf of bone seen in *H. erectus*. Although *H. heidelbergensis* was named after a city in Europe, it has also been found in Africa and East Asia. The species is represented by fossils ranging in age from about 200,000 to 800,000 years ago, although there is still debate whether this assemblage represents more than one species.

By 200,000 years ago, human evolution gets a bit more complicated. In Europe, the Middle East, and parts of central Asia we see evidence of a form known as the Neandertals, a group of archaic humans that appear to have evolved from *Homo heidelbergensis* in Europe and persisted until about 28,000 years ago. There has been considerable debate over whether the Neandertals represent a separate species or are a variant of *Homo sapiens*, a subject of a later myth (Myth 33). Like *H. heidelbergensis*, the Neandertals had a long and low skull, but had an even larger brain size. They also had a number of unique anatomical features, such as a protrusion on the back of the skull (known as the occipital bun), swept-back cheekbones, large incisor teeth, and a large nose and mid-facial region (Figure 3.6).

At the same time the Neandertals lived in western Eurasia, populations of modern *Homo sapiens* appeared 200,000 years ago in Africa, perhaps the descendants of some population of African *H. heidelbergensis*.

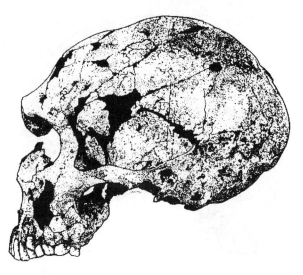

Figure 3.6 Side view of Neandertal, La Ferrassie specimen, France. Source: Cartmill and Smith, 2009. Reproduced with permission of Wiley and Matt Cartmill.

By 100,000 years ago, populations of modern humans began dispersing from Africa and spreading throughout Eurasia and beyond, ultimately reaching even Australia and the New World (Figure 3.7). The dispersion and the interaction of modern humans with preexisting humans such as the Neandertals will be discussed in a number of later myths. For now, we note the major evolutionary changes from *H. heidelbergensis*—a further increase in both absolute and relative brain size, a shift in the shape of the skull to a more well-rounded basketball shape often with a vertical forehead, further reduction of the face and teeth, and the development of a prominent chin (not seen to any great extent in earlier humans).

Given this brief (and somewhat oversimplified) evolutionary review, we now turn to the evolution of stone tool technology. The earliest stone tools date back to 3.3 million years ago in East Africa and appear to represent the beginning of stone tool use.[17] These tools are named Lomekwian tools, after a site in Kenya, Africa, and it is not clear who made them. The date is too early for even the oldest-known fossils of *Homo*, so it seems more likely that they were made by *Australopithecus* (but which species?). Better known early stone tools are found at about 2.5 million years ago in East Africa and consist of stone cores from which a few flakes have been removed through direct striking of one stone (the core) with another (the hammer). These tools are named Oldowan tools after a site in

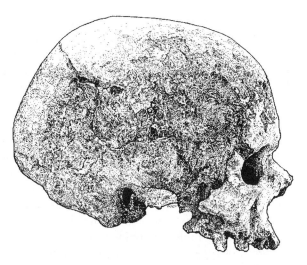

Figure 3.7 Side view of Cro-Magnon, prehistoric *Homo sapiens* specimen, France. Source: Cartmill and Smith, 2009. Reproduced with permission of Wiley and Matt Cartmill.

Tanzania, Africa, and are rather crude compared to later stone tools. The removal of a few flakes give these tools a cutting edge that can be useful in a variety of tasks including disarticulating a carcass or breaking bone. (Although useful, I would not want to shave with one!) Archaeologists have also noted that the flakes of stone removed during the creation of an Oldowan tool (the so-called waste flakes) are razor-sharp and were likely used for cutting, such as the removal of flesh from a carcass. These flakes made up part of the overall Oldowan toolkit.

Oldowan tools are found in association with *Homo habilis* and *Homo rudolfensis* in great abundance. Although simple, these tools allowed our ancestors to obtain food that would otherwise be unavailable and the main activity was likely scavenging of dead or dying animals. Humans, past or present, lack the physical ability to cut through an animal hide or to crack open large animal bones. We do not have the sharp teeth or claws found in carnivores and have to make up for this lack by inventing sharp tools. A stone core with a sharp edge or a razor-sharp flake would make all the difference in survival.

Homo erectus also used Oldowan tools, although later forms showed some improvement in technique. However, *H. erectus* went further and invented a new type of stone tool tradition known as the Acheulean (named after a site in Europe and sometimes spelled "Acheulian"). Acheulean tools use a bifacial method of manufacture where stone flakes are removed from both sides of the tool. This method creates a stone tool that is more symmetric than an Oldowan tool, and one that has a recognizable shape. Smaller flakes need to be removed; archaeologists have shown that the toolmakers would have needed to use softer, more resilient materials such as bone and antler to remove these flakes carefully. The most common form of Acheulean tool was the "hand axe," a rather large tool that can range in size from 15 to 20 centimeters (6 to 8 inches) in length. These hand axes were not used as axes, but could be used for a variety of tasks including butchering, scraping hides, breaking bones, chopping wood, and digging in the ground. All in all, Acheulean hand axes appear to be general-purpose tools.

Acheulean tools first appear in Africa about 1.8 million years ago.[18] However, *H. erectus* populations in Asia did not have Acheulean hand axes, but instead had less complex chopping tools. The geographic difference in *H. erectus* tool types was first noted by archaeologist Hallam Movius in the 1940s; hand axes tend to be found in the western part of the Old World and chopping tools in the eastern part. Perhaps there were toolmaking differences among African *H. erectus* populations and the ones who moved out of Africa did not have this skill or lost it over time. Another possibility is that Asian *H. erectus* populations used bamboo as a raw material for many tools, including knives and scrapers, so that sophisticated stone tools were just not needed. In any case, even though *H. erectus* invented Acheulean tools, not all *H. erectus* populations used them, and they did not abandon Oldowan tools right away.

Acheulean tools are also interesting because they represent the longest existing type of stone tool, found not only with *Homo erectus*, but also with populations of *Homo heidelbergensis*. For many hundreds of thousands of years, *H. heidelbergensis* continued to use Acheulean tools in Africa and Europe. It is not until about 300,000 years ago that *H. heidelbergensis* invents a new type of stone tool manufacture known as the Levallois technique. These Levallois tools are also known as preparedcore tools, which is an apt description of how they are made. The toolmaker prepares the initial core by removing small flakes around the perimeter of the stone core to get a particular shape. Small flakes are then removed from the front surface of the core and then a sharp blow is used to strike off the finished tool from this surface. The surface is then reflaked and a second blow is used to strike off another tool identical to the first. This is essentially the start of mass production, allowing a number of identical tools to be made from a single core. It is also an efficient method because you can get more mileage out of a single stone core. Because it is hard work to move stones from where they are found to where you need to use them, the more use you can get out of individual stone cores, the better. Making these tools requires considerable skill and foresight to be able to see the final product emerge from a number of intermediate steps. It is worth noting that Neandertals, who are descended from *Homo heidelbergensis*, also used tools that were made using the Levallois technique.

We often refer to the early prehistory of stone tool technology as the Old Stone Age or the Paleolithic (which literally means "old stone"). The Oldowan and Acheulean traditions are typically labeled as Lower Paleolithic tools and the prepared-core tools as Middle Paleolithic. The term Upper Paleolithic describes the diverse Stone Age cultures found in association with prehistoric *Homo sapiens* and include further sophistication in methods of manufacture, use of bone as a resource, and an increased level of technological complexity. Overall, we do see a pattern of change throughout the Paleolithic ranging from the relatively simple tools of the Lomekwian and Oldowan to the diverse and sophisticated tools in the Upper Paleolithic. Although this happens at the same time that we see a transition from early *Homo* and *Homo erectus* to *Homo heidelbergensis* to *Homo sapiens*, these biological and cultural changes do not occur in lockstep.

Oldowan tools were used by *Homo habilis*, *Homo rudolfensis*, and *Homo erectus*. Acheulean tools were invented by *H. erectus*, but not all populations used them, and Acheulean tools continued to be used for at least half a million years by *Homo heidelbergensis*. Although *H. heidelbergensis* did invent the Levallois technique, it is also used by Neandertals and by early *Homo sapiens*, who did not develop Upper Paleolithic tools until about 50,000 to 100,000 years ago.

There appears to be a lag between the appearance of a new species and a new type of stone tool technology. In one sense, this lag should not be surprising, as it would be rather strange to see complete congruence between biological and cultural change because that would place toolmaking in the realm of something that could be tied to a specific set of genetic changes. Instead, what might be changing is basic capability. Why, for example, were Acheulean tools used for so long by *Homo erectus*? Were they sufficient for their needs or was there some sort of limitation on their abilities to create and use technology? If so, then why did it take *Homo heidelbergensis* so long to invent a new type of tool? Why did the emergence of *Homo sapiens* eventually give rise to a rapid increase in new types of technologies, both in prehistoric times and after (and continuing today)? Why did these changes not occur when *Homo sapiens* first appeared 200,000 years ago? How do we explain both changes and lags? We are likely seeing a complex mix of changes in capability, environmental demands and limitations, and changing climate, among many other factors, all of which likely had difference levels of influence at different times. In any case, we see that there is not a one-to-one correspondence between species and tool types. Finding a particular type of tool, such as Acheulean tools, without other clues cannot tell use exactly who the toolmaker was in many cases.

Myth #25 *Homo habilis* definitely made shelter

Status: Homo habilis *has been found in the lower levels of Olduvai Gorge in Tanzania, East Africa. One site containing remains of stone tools and animal bones also has a group of rocks that appears to be arranged in a circle over 4 meters (13 feet) in diameter. This stone circle has been interpreted as an early form of shelter, perhaps a windbreak or the remains of a hut. Although this is possible, no similar structures have been found associated with* Homo habilis *and it is likely that the arrangement of rocks was due to natural processes, such as the growth of a tree through the ground.*

When we interpret the fossil and archaeological evidence, our interpretations are often not firm conclusions, but are instead hypotheses that need to be confirmed or rejected with further data and analysis. Sometimes the difference between a hypothesis and a conclusion can get lost in the shuffle, particularly if the conclusion fits the available evidence in accord with our own predilections. In the case of human evolution, our biases might lead us to view early ancestors as being more or less like us than might actually be the case. This is particularly a problem when interpreting remains of past behavior. For example, animal bones showing signs of butchering indicate that our ancestors used stone tools to process animal flesh, which might be due to hunting, but also could be due to scavenging.

Another example to consider is burial of the dead. Neandertals and *Homo sapiens* deliberately buried their dead but earlier species did not. Archaeologists found fossil pollen, corresponding to spring wild flowers, in one Neandertal burial. How can we interpret the presence of pollen in the grave? Using ourselves as a model, we note that we commonly place flowers in or on a grave and then extrapolate this behavior back into the past. Is this valid? Perhaps, but there might also be alternative explanations. The pollen could have entered the grave through the actions of burrowing rodents whose fur carried pollen. More analysis would be needed to distinguish between different hypotheses. The point is that our first interpretation, based on the behavior of modern humans, might be in error and we need to always remember that although our ancestors may have been like us in some ways, they might have also been different.

Interpretative bias can be a problem when dealing with the early members of the genus *Homo*, such as *Homo habilis*. Their placement in the genus *Homo* is based on an increased brain size and stone tools. The idea of a larger-brained biped with technology sounds very human to us and it will be tempting to complete the rest of the picture by seeing evidence of specific human behaviors, such as big-game hunting, use of fire, and the construction of shelter. However, *Homo habilis* was not a modern human (and, indeed, some paleoanthropologists argue that based on primitive features of the skeleton, it does not really belong in the genus *Homo*).[19] The presence of stone tool technology does not automatically mean that other human behaviors were present.

An example of alternative explanations comes from Olduvai Gorge in Tanzania. Olduvai Gorge lies within the Great Rift Valley in East Africa, a long cleft in the earth caused by the movement of continents. Olduvai Gorge is a canyon that has been formed by an ancient river. The exposed sides of the gorge provide a view into the last 2 million years of earth's history. The famed anthropologists Louis and Mary Leakey started working at Olduvai Gorge in the 1930s looking for signs of early human evolution. The Leakeys were rewarded by considerable evidence of early stone tools, the Oldowan tradition. Once they found the tools, a major focus of their work was to find who made the tools. In 1960, they finally found the toolmaker, with the first remains of what would soon be named *Homo habilis*. The increased brain size (relative to *Australopithecus*) and the presence of stone tools convinced Louis Leakey and colleagues to place this new species in the genus *Homo*.

Olduvai Gorge is made up of seven different geological strata, or beds. The oldest level, Bed I, lies on top of a layer of volcanic rock and dates to between 1.75 and 2 million years ago. Many of the original *Homo habilis* fossils as well as Oldowan tools have been found in Bed I. The DK site in the lower part of Bed I is particularly interesting; in addition to numerous Oldowan tools and animal bone fragments, there are a number of chunks of lava roughly 10 to 25 centimeters (4 to 10 inches) that were found lying in a circle 4 to 5 meters (13 to 16 feet) in diameter.[20] The circular arrangement suggests that these rocks were placed there on purpose, perhaps as a windbreak or as part of a more complex shelter. Mary Leakey suggested that the circle strongly resembled the kinds of stone circles left when living hunters and gatherers build shelters. Branches are placed in the ground in a circle and then bent toward each other, forming a wooden framework for a hut. Stones are placed next to the branches to help anchor them in the ground.[21] In a short time, the branches would decompose leaving only the stones behind, which is what we would see today.

The presence of a shelter at this site suggested to some that this was an example of a home base to which *Homo habilis* returned with food that was presumably hunted. Butchering would have taken place at this site and shared among the members of the group. The accumulation of animal bones and stone fragments represented a long period of occupation by the hominins living there. Because living hunters and gatherers perform most activities around their home bases, we see that debris accumulates in some places. Extending this observation into the past led to the conclusion that a concentration of stones and animal bones at Olduvai Gorge (and elsewhere) was evidence of home bases. Taken all together, the evidence suggests a fascinating picture of the life of our early ancestors, and very "human" in behaviors such as hunting, food sharing, and construction of shelter.

Although it is possible that the stone circle at the DK site is a windbreak or the remains of a hut, we cannot know for sure, as there are also natural processes that could account for the patterns that we see. Archaeologist Richard Potts has investigated the stone circle and suggested that it, as well as other evidence suggesting that the DK site was a home base, is due to natural causes.[22] The stones are all from the underlying level of lava at the site, and could have been produced by the spread of roots radiating outwards when a tree grew up out of the ground. When this happens, the rock will split and spread out in a circular arrangement, producing the kind of pattern that we see at the DK site. Potts also notes that fragments of bone and stone found inside the stone circle could have been deposited there by running water. Indeed, the bones and stones throughout the entire DK site show evidence of abrasion due to water.

Did *Homo habilis* make shelter? It is possible, but the work of Potts and others shows that there is no definite evidence that they did so, and the suggested evidence can be explained in terms of natural processes. Therefore, to extrapolate from the behavior of living humans and infer similar behaviors in early *Homo* is premature. What about for later species of *Homo*? Our ancestors would have certainly needed shelter when they moved out of Africa into more temperate (and colder) climates. We have evidence that *Homo erectus*, as well as later humans, used caves for shelter, but what evidence is there for constructed shelters? Anthropologist Richard Klein notes that shelter would have been necessary for human ancestors when they moved into Eurasia,[23] and there are no caves or natural shelters available in the open-air sites we have found. However, he notes that the actual evidence for constructed shelter is not firm prior to about 130,000 years ago.

As we continue searching for evidence on the lifestyle of our ancestors, we need to guard against the trap of seeing more in the evidence than might exist and to be skeptical about early evidence of modern behaviors. As Potts notes, "Although it is a great temptation to see signs of modern humanity in the activities of these early ancestors, the evolution of humans must be seen from a perspective of change. The study of human evolution requires the ability to see antecedents, things that were not quite as they are today or in recent times."[24]

Myth 26 Our ancestors have always made fire

Status: Control of fire is a significant achievement in human evolution, but not all of our ancestors used or made fire. There is no evidence of using fire for early hominins such as Australopithecus. *It seems unlikely that* Homo habilis *used fire. The oldest definite evidence of the use of fire is from Israel 790,000 years ago and perhaps from South Africa a million years ago. There is evidence that* Homo erectus *in Asia used fire, although not as extensively as once thought. Widespread use of hearths only appears in the last 350,000 years.*

Several key events stand out in the history of our species. Working back from the present, these include the beginnings of space exploration, the agricultural revolution, the invention of art, and the origin of stone tool technology. Another key event, a bit harder to date, is the use of fire. There is something primal about staring into a fire, imagining (from the comfort of our living rooms or camp sites) what life must have been like when much of our lives depended on fire, and what life must have been like before our ancestors had the ability to harness fire.

Being able to make and/or control fire is certainly an important event in our evolution that transformed the lives of our ancestors. Fire represents the first conversion of matter into energy and the first control of non-kinetic energy. Fire can keep you warm, something that would be critical for our ancestors, particularly as they began expanding out of the tropics. Fire provides a means of cooking, which can be very adaptive. Fire also provides light, which can scare away animals at night and extend our days. With this new form of light, our ancestors could be productive after sunset, working on tools and weapons, planning the next day's events, or just socializing. The control of fire marks the beginning of freeing ourselves from the natural cycles of the world around us.

Is the use of fire an ancient human behavior? A common image of our ancestors is that of the "cave man," with a family unit gathered about an open hearth near the entrance of a cave, stone tools and spears close at hand, and a large haunch of animal flesh ready to barbeque. This image pervades literature and popular culture, and we often consider the control of fire a very ancient accomplishment originating deep in prehistory. However, which of our ancestors actually could use (and make) fire?

Evidence for the control of fire comes from several sources, including the presence of carbon (including charcoal) and the accumulation of ash in hearths. Use of fire is also detected from burned bones, which makes sense if our ancestors were cooking over the fire. However, it is not a simple task to provide definitive evidence of controlled fire. Some evidence might not be available; for example, charcoal deteriorates quickly in the open air due to water and wind erosion. In addition, we have to distinguish between controlled fire and fire that occurs naturally, such as brush fires due to heat or fire caused by lightning strikes. The best evidence for controlled fire is when we find burned animal bones and flints as well as charcoal in an enclosed area, such as a cave, and in the absence of any evidence of flowing water in the cave, which could make give a false impression of association.[25]

A further complication is that the evidence that our ancestors controlled fire does not necessarily mean they were able to *make* fire. We can easily imagine a situation where some early humans came across a downed branch burning after a lightning strike and then took it back to their cave and used it until it eventually went out. Did our ancestors discover how to make and use fire early on, or did they use fire intermittently when available from nature?

The date for the oldest *use* of fire in human evolution is unclear and the debate over what constitutes the earliest evidence is somewhat controversial. It is clear, however, that control of fire did not occur before the emergence of the genus *Homo*, and more specifically *Homo erectus*. Although Raymond Dart had first suggested the use of fire for *Australopithecus* based on blackened animal bones, this discoloration was instead due to chemical staining (as discussed in Myth 17).

The use of fire is associated with the genus *Homo*, although there is considerable debate over the earliest date. There have been claims for ancient use of fire at several sites in Africa, the oldest of which dates to about 1.6 million years ago. This evidence, consisting of burned clay, might instead reflect the action of a naturally smoldering fire. Likewise, evidence of burned bone from South Africa dating back between 1 and 1.5 million years ago could also be explained by natural causes.[26] Recent suggestive evidence from Wonderwerk Cave in South Africa, a site with Acheulean tools, dates to 1 million years ago and consists of burned bone fragments and the remains of plant ash.[27]

The most widely accepted earliest confirmed date for the use of fire comes from the site of Gehser Benot Ya'aqov (GBY) in Israel, dating to almost 790,000 years ago.[28] Here, researchers found burned pieces of flint, as well as burned fruits and wood. When flint is burned at high temperature it leaves characteristic fractures that can be identified. These burned artifacts were not distributed uniformly throughout the site, but instead occurred in several clusters, which were suggested to have corresponded to the location of hearths. Although wildfires can reach high enough temperatures to burn flint, such fires would be widespread and not clustered. At the GBY site, less than 2 percent of the flint and wood was actually burned, too low for a wildfire but consistent with a controlled fire.

There are Acheulean tools at the GBY site but no hominin bones, which makes it difficult to identify who used the fire. Acheulean tools are associated with both *Homo erectus* and *Homo heidelbergensis* (see Myth 24). The date of 790,000 years ago fits with *H. erectus*, but also with early *H. heidelbergensis*. To date, *H. erectus* has not been found conclusively further west in Eurasia than the Dmanisi site in Georgia, while *H. heidelbergensis* did live as far west as Spain and England. Perhaps the GBY site marks the movement of early *H. heidelbergensis* populations out of Africa and through the Middle East into Europe. On the other hand, western expansion of *H. erectus* into the Middle East is not unreasonable. We will need hominin fossils from the area to provide more insight into the specific ancestor, but for the moment, we can see that the GBY site gives strong evidence for controlled fire by some species of *Homo*.

Association of a hominin species with controlled fire is found in East Asia at the famous cave site of Zhoukoudian, near Beijing, China. Excavations in the 1920s and 1930s revealed fossils of "Peking Man," now formally assigned to the species *Homo erectus*, who lived there between 400,000 and 780,000 years ago. Early discoveries at Zhoukoudian were instrumental in giving rise to a popular notion of the lives of "cavemen" including evidence of tool use, hunting, and the use of fire. A number of these early ideas now seem a bit exaggerated.

There is still suggestive evidence that *Homo erectus* used fire at Zhoukoudian but it is not as strong a case as was claimed originally.

At the time of discovery, the evidence did point to use of fire, including carbon and ash deposits and burned stone tools and animal bones. Some of these points of evidence have been challenged. For example, the carbon deposits are more likely due to sediments left by water flowing into the cave.[29] Further analysis showed no charcoal and the presence of certain phosphate minerals turned out not to be from ash after all. There is an association of tools made from quartz and burned animal bones, but it is possible that this association resulted from the flow of water into the cave, which means we cannot pinpoint the locations of ancient campfires as once thought.[30] Although the evidence does support some use of fire by *Homo erectus*, the long-standing view of frequent hearths in the caves does not appear to be the case. Use of fire at Zhoukoudian might not have been as frequent as we once thought.

It has been suggested that early hominins in Africa may have used natural fire on occasion but did not actually create fire. In Africa, the frequent occurrence of lightning strikes would have created many natural fires that our ancestors would use when afforded the opportunity. Likewise, *Homo erectus* at Zhoukoudian might have used naturally occurring fire on occasion. The regular use of fire, at least in the Middle East and Europe, took place about 350,000 years ago. The clearest evidence of habitual fire use comes from the analysis of the frequency of burned flints in different layers of occupation at Tabun Cave in Israel. Here, fire was rare in early times, but became frequent between 357,000 and 324,000 years ago.[31] These dates are consistent with similar evidence in Europe, pointing to a relatively recent spread of habitual fire use. The relatively late origin of controlled fire causes us to reconsider a common assumption that the earliest hominins dispersing into Eurasia must have had fire to adapt to the colder environments outside of the tropics. Although it seems clear that humans used fire increasingly after 300,000 years ago, in part to adapt to extremely cold climates in glacial Europe, the initial colonization of much of Eurasia occurred before evidence of widespread use of fire.[32]

Fire is certainly a major adaptation in human evolution, but it may not be as ancient as once thought, at least in terms of frequent use and the ability to create fire rather than to rely on natural fires. It seems likely that both *Homo erectus* and *Homo heidelbergensis* used fire on occasion, but it was not until the later days of *H. heidelbergensis* that fire became a significant and frequent part of our ancestor's lives. By the time the Neandertals and modern *Homo sapiens* had appeared, fire took on the important role that some once thought began much earlier in time.

Myth 27 Early humans got all of their meat from hunting

Status: The association between stone tools and butchered animal bones at Homo erectus *sites has long been taken as evidence of hunting, particularly big-game hunting. Close analysis shows that a considerable amount of animal protein may have been obtained by scavenging dead animals and not by hunting. The relative amount of scavenging versus hunting activity in early humans continues to be debated. Evidence of significant hunting is more clearly associated with later species in the genus* Homo.

Hunting has long been considered a unique feature of early human evolution associated with the origin and development of stone tool technology. *Homo erectus* in particular has often been portrayed as an active hunter, including big-game hunting. In addition to stone tools, we find many butchered animal bones at *H. erectus* sites. At the famous Zhoukoudian site in China, archaeologists have found remains of thousands of deer, in addition to other species of medium to large animals, including horses, buffalo, and rhinos.[33] These remains, along with the remains of *Homo erectus*, stone tools, and suggested evidence of fire, all added up to an accurate description of the life of early humans and contributed to the popular images of cave men. Both biological and technological changes in the evolution of *Homo* could be seen as a reflection of the needs and benefits of a meat-rich diet. Hunting became viewed as the driving force in human evolution, a view often referred to as the "man the hunter" model. The next step was to take living hunters and gatherers as representative of the behaviors likely to be exhibited by early humans such as *Homo erectus*.

The idea that *Homo erectus* was a habitual big-game hunter has been questioned over the last 30 years. Some have questioned the relevance of an association of animal bones with remains of *H. erectus*, noting that both may have accumulated at the same site due to natural processes. However, there is sufficient evidence that early humans such as *Homo erectus* butchered animal bones. How do we know this?

One of the most important questions about animal bones is figuring out what happened to them before we find them. Some of the best information comes from looking closely at the animal bones under extreme magnification using a scanning electron microscope (SEM). For example, when an animal dies and its bones are later stepped on, this will leave a series of shallow scratches on the animal bone. When a carnivore kills and eats an animal, its teeth will make a distinctive mark on the animal bone. Experimental studies show that carnivore gnawing produces broad "U-shaped" marks with pits made by the canine teeth. Other patterns of scratches come from rodents gnawing on bones. When a human uses a stone tool to remove meat from a bone, it too leaves a characteristic signature: a sharp, narrow, and deep V-shaped groove with a series of parallel scratches within. Stone tool cut marks tend to cluster in sets at the site where a human was butchering, such as disarticulating a joint or removing meat from a limb bone.[34]

The presence of stone tool cut marks shows us definitively that early humans such as *Homo erectus* did butcher animals. This fits with other observations that *H. erectus* had a significant amount of meat in its diet, such as a body shape consistent with a reduction in the size of the intestines, possible with a new diet (see Myth 22). The big question, however, is *where* did they get the meat? A reasonable answer would be hunting, but is this the only possibility? No, because one could also obtain animal flesh by scavenging and removing the flesh from recently deceased animals, who in turn could have died of natural causes or killed and partially eaten by carnivores. Another possibility would be chasing away the predators after a kill, in which case our ancestors would still be scavenging, but would have primary access to the kill.

When we come across the remains of an animal, SEM analysis can help us figure out what happened. If we find a bone or bone fragment with stone tool marks, we can determine that some butchering occurred. What if we find a bone with both stone tool marks and the teeth marks of another animal? There are two general scenarios to explain the presence of both tool marks and teeth marks on the same bone. In the first case, we can imagine that early humans were the hunters, who killed and then butchered and ate part of the kill, and then scavengers, such as hyenas, munched on the leftovers. The second scenario reverses the role of humans and nonhumans. Here, we imagine that a carnivore kills the prey, eats part of it, and then leaves, or perhaps is chased off by humans. Our ancestors then come in and use stone tools to scavenge the remaining flesh. In the first scenario, humans are hunters, and in the second, they are scavengers.

How can we distinguish between these two different explanations of finding both stone tool cut marks and tooth marks on animal bones? The answer is that in some cases, the stone tool cut marks and the tooth marks overlap, and we can then tell which one came first. As analogy, consider two cars driving through an intersection of two muddy dirt roads.

The first car that comes through the intersection puts down a set of tracks. The second car then comes through on the other road and its tracks are laid down on top of the first set of tracks. If this were to happen, you would easily be able to tell which set of tracks came first. In the case of the animal bones, microscopic examination allows us to tell whether the stone tool cut marks came first or if the dental marks came first. If the stone tool cut marks came first, then humans were doing the initial butchery, and the tooth marks likely represent scavenging of the leftovers by other animals. If the prey were killed first by a carnivore and humans then used stone tools to scavenge the remains, then the tooth marks would come first.

It turns out that there are many cases in early human evolution where the tooth marks came first and the stone tool cut marks came afterwards. Paleoanthropologists Noel Boaz and Russell Ciochon note that this is the case for the animal bone remains at Zhoukoudian.[35] In such cases, *Homo erectus* was a scavenger and not a hunter of big game. According to some, this is not unexpected, as the stone tool culture associated with *H. erectus* does not include evidence of the kind of weapons that would be useful in hunting larger game animals, such as spears. This does not mean that *H. erectus* never hunted, as it would have been within their ability to hunt small game. In addition, we should not look at the question of meat eating as being *either* scavenging or hunting, as both could have been important. Even a habitual hunter would still use scavenging when appropriate. (Would you pass up the chance to scare off predators from a fresh kill if you were hungry?) Although debate continues as to the relative importance of scavenging and hunting for *H. erectus*, it seems clear that the image of a full-time habitual big-game hunter with all the proficiency of living hunter-gathers seems outdated.

There is also evidence that hunting ability and technology increased during the course of human evolution. According to some anthropologists, *Homo heidelbergensis* is the first to be considered a habitual big-game hunter.[36] At the Gran Dolina site in Spain, dating back 800,000 years ago, there are remains of many small and large mammals that have been butchered. These bones show signs of butchering and tooth marks, and in all the cases where the two overlap, the stone tool cut marks came first.[37] The stone tool cut marks are also found on bones throughout the animal, including parts that would have contained a great deal of meat. This pattern supports the idea that early humans had priority access to the meat,[38] because if they were scavengers the cut marks would occur most often in parts of the bones that were left over by carnivores. Although some of this evidence could also be consistent with scavenging of complete carcasses after chasing away a predator, the bulk of the evidence is more consistent with humans doing the hunting. (I refer to the hominins at this site as *Homo heidelbergensis*, although some prefer to assign them to a separate species, *Homo antecessor*, which is transitional between *H. erectus* and *H. heidelbergensis*.)

Evidence of improved hunting technology comes from 400,000-year-old archaeological sites in the town of Schöningen in Germany. Although no hominin fossils were found at the site, the location and age of the site suggests that *Homo heidelbergensis* was present. One of the sites contained flint tools, evidence of fire, and the remains of 19 horses, including many bones that show stone tool cut marks associated with butchering. The distribution of the butchered horses suggests that a planned hunt of an entire herd at one time. Incredibly, more than half a dozen wooden spears were found at the site.[39] Typically, wooden tools and weapons decompose quickly, but there are rare cases where they can be preserved in sediments. The wooden spears at Schöningen are roughly 1.8 to 2.5 meters (6 to 8 feet) in length. The spears were made from spruce or pine, and each shows signs of having been cut and shaped from small trees. The balance of the spears has the maximum thickness and weight about one third of the distance from the tip, which is similar to that of a modern-day javelin. Because of the problem of preservation, it is possible that hunting with wooden spears occurred before this time but we have no direct evidence. However, the site does establish that hunting spears had been used *at least* 400,000 years ago.

More evidence of hunting big game with spears comes from the 300,000-year-old Boxgrove site in England.[40] A human tibia was found at this site and, based on the age and location of the site, possibly represents *Homo heidelbergensis*. Stone tools have been found at the site, as well as the butchered remains of a number of large mammals, including deer, horses, bison, and rhinos. Stone tool cut marks are found throughout the bodies, and where they overlap with tooth marks, the stone tool marks always came first. Although this pattern could be due to scavenging after humans chased away a predator (another form of primary access to the meat), the mammal remains include a large rhino that probably did not have any natural predators, which suggests that humans were the hunters. The possibility that the rhino had died of natural causes and was then scavenged by humans can also be rejected because there is not sign of illness in the rhino's bones. Another interesting find at the Boxgrove site is a shoulder bone from a horse that shows a wound that would have been caused by a wooden spear. Taken together, the site shows strong evidence that early humans at Boxgrove were proficient hunters by this point in time.

This does not mean that there was no scavenging, but instead suggests that active big-game hunting had developed fully by this time.

One of the things we need to keep in mind when looking at the evidence for early hunting activity is the danger involved in using wooden spears, as shown in an interesting analysis of trauma among the Neandertals of Europe, descendants of *Homo heidelbergensis*. Some stone projectile points have been found of a type that could have been hafted onto wooden spears, creating a formidable weapon. However, these spears would have been too heavy to throw with any serious amount of force; instead, they were likely used as thrusting weapons. Although this will work, using such weapons requires you to be up close to the animal, which can be very dangerous to say the least! This danger is reflected clearly in Neandertal skeletons, which show a pattern of neck and head trauma similar to a group of living humans who also work up close with large, strong, and dangerous animals—rodeo riders.[41]

Over time, further changes in hunting technology were made by *Homo sapiens* that allowed hunting at a distance using early projective weapons.[42] One such weapon is the atlatl, which is a bone or wooden spear-throwing device into which a spear is rested. When using an atlatl the throwing force is increased, making it useful to hunt animals without getting too close. Atlatls have been found in both the Old World and the New World (such as the Aztecs). The oldest evidence for atlatls is from France some 18,000 years ago. Another type of projective weapon is the bow and arrow, which dates back at least 20,000 years ago in Africa and Europe. Long-distance weapons would clearly reduce personal danger.

The archaeological record shows that humans became more sophisticated and adept hunters over time, increasing their ability to acquire animal protein through hunting. However, the available evidence suggests that extensive big-game hunting may not have been present right from the start with *Homo erectus*. Like so many other features of human evolution, extensive big-game hunting developed over time.

Species with larger brow ridges are more ape-like

Status: In the nineteenth century, large brow ridges were considered a primitive ape-like characteristic and the relationship of human fossils to living humans was interpreted in part based on the size of their brow ridges. Evolution of bow ridge size is actually more complex and during the course of the evolution of the genus Homo *brow ridges have increased and then decreased in average size over time. A number of models have been proposed to explain the presence of large brow ridges in past species such as* Homo erectus. Homo heidelbergensis, *and the Neandertals. To date, the best explanation is biomechanical stress.*

When students in my paleoanthropology class examine fossil casts of *Homo erectus*, they are immediately impressed by the size of the brow ridges in some of the specimen. (A particularly impressive specimen is the cast of the OH 9 fossil skull cap from Olduvai Gorge; students often write "Huge!!!" with multiple exclamation points when completing their lab assignments.) My students also see that although most *H. erectus* specimens in the lab have larger brow ridges than in living humans, there is variation, and many ask why some skulls show smaller or larger brow ridges. The curiosity about brow ridges continues into later labs, as students examine casts of *Homo heidelbergensis* and Neandertals, noting that they still have impressive brow ridges, although somewhat different in form.

The fascination with brow ridges has long been a part of the general interest in human evolution. As author and anthropologist Mary Doria Russell has noted, the brow ridge was once described by anatomist Hermann Schaafhausen as "a most remarkable peculiarity."[43] People have read many things into the meaning of large brow ridges, both in terms of evolutionary adaptations and in terms of the mental capacities of our ancestors.

The brow ridge is more technically known as the *supraorbital torus*. A torus is a ridge of bone, and "supra" is Latin for "above," making the supraorbital torus the ridge of bone above the eye orbits. Brow ridges vary quite a bit among living humans but seldom form the same type of complete torus seen in earlier species,[44] and they are much smaller on average than in apes or in most earlier species of *Homo*. Today, men on average have larger brow ridges than women. In fact, the size of the brow ridge is one of the cranial traits used by forensic anthropologists when determining sex from skeletal remains. There is also variation between populations in brow ridge size. Natives of Australia and Melanesia tend to have rather large brow ridges on average, as do some natives of South America, while people of European ancestry have smaller brow ridges, and people of African and East Asian ancestry even smaller.[45]

A large brow ridge has often been viewed as a primitive, ape-like trait. Among living primates, large brow ridges are found in African apes but not in Asian apes, and in only a few monkey species. Brow ridges in apes and humans were discussed in the nineteenth century as scholars debated whether Darwin's idea of evolution was correct, and whether apes and humans were truly related to each other. As part of this debate, paleontologist Richard Owen (best known for first naming dinosaurs) was a critic of Darwin's work and argued against a common ancestry of apes and humans by using the development of brow ridges as an example. Owen stated that the brow ridge has no functional significance as no muscles connect to it and therefore would not be subject to natural selection. In his view, the presence or absence of a brow ridge reflects common ancestry. He then argued that because apes had prominent brow ridges, if apes and humans shared a common ancestor, then humans should also have prominent brow ridges (and not the much-reduced form we see in living humans). According to Owen, the absence of larger brow ridges in living humans is proof that apes and humans did not share a common ancestor.[46]

Keep in mind that there were no human fossils (other than recent modern humans) known at the time that Owen made this argument. Shortly afterwards, the first Neandertal specimen was announced and although it had a large cranial capacity as did modern humans, it had a low and long skull with very noticeable brow ridges. If one accepted the Neandertals as some form of human, it became clear that Owen's view that a prominent brow ridge was exclusively found in apes could not be true. However, the association of large brow ridges with ape-like qualities persisted. As writer John Reader notes, from this point onwards in history, large brow ridges "have become symbols of the prehistoric human form."[47]

The existence of primitive and supposed ape-like features of Neandertals led many to focus on the anatomical differences between Neandertals and modern humans. William King, the anatomist who first designated Neandertals as a different species—*Homo neanderthalensis*—noted that the large brow ridges, as well as other features of the skull, showed Neandertals to be closer to apes than to living humans. King speculated further on the mental abilities and moral characteristics of Neandertals, stating that he felt "constrained to believe that the thoughts and desires which once dwelt within it never soared beyond those of the brute."[48] He went on to compare the Neandertal with the Andaman Islanders (who live on islands south of India and Myanmar), by noting that "The Andamaner, it is indisputable, possesses but the dimmest conceptions of the existence of the Creator of the Universe: his ideas on this subject, and on his own moral obligations, place him very little above animals of marked sagacity."[49] Even so, he noted that Andaman Islanders belonged to *Homo sapiens*, but this was not true of Neandertals. The racist tone characteristic of the times is apparent, as is the inherent ranking of fossil and living humans along a ladder of progress from ape to human, where the Andaman Islanders were placed on the lowest rung of the ladder in living humans, with Neandertals being the link to apes. The designation of Neandertals as ape-like did not rest entirely on evaluation of the large brow ridges, but they certainly played a large part of the anatomical interpretation of Neandertals as "savages."[50]

If we extend this idea of using brow ridges as a link between African apes and living humans, we might expect that we would see a gradual reduction in the size and prominence of brow ridges during the evolution of the genus *Homo* as we move further away from an ape-like ancestor. Actually, this is not the case. *Homo habilis* had relatively small brow ridges, followed by an increase in size for *Homo erectus*. Large brows continue in *Homo heidelbergensis* and the Neandertals, but shift in form from a long continuous shelf of bone to brow ridges that are more arched and less thick in the middle. Brow ridge size is smaller in modern humans. The evolution of brow ridges is not a simple trend, but rather one where a trait undergoes a reversal in size. The size of brow ridges in any given species is therefore not an index of how similar they are to apes, and it cannot be used to rank or sort species on a family tree.

This conclusion raises other questions. Why do some species have larger brow ridges? What is the function of large brow ridges? A number of interesting ideas have been proposed to explain the large brow ridges of Neandertals (as well as *Homo erectus*).[51] There were some early suggestions that large brow ridges were simply the result of disease, an idea that soon was rejected when evidence accumulated that entire samples of fossils had large brow ridges, as it is not possible that everyone in a population had the same pathology. Some have proposed that brow ridges served to threaten or intimidate other members of the group. Although this might be a possible consequence of large brow ridges, it is not clear how variation in brow ridge size would translate into actual difference in survival and reproduction in such a way as to allow selection for larger brow ridges to occur over time, or how this would relate to population differences.

Some explanations for large brow ridges have focused on protection of the eyes from the sun and rain, where large brow ridges served as a visor of some sorts. I am not sure if a larger brow ridge provides much value. Try holding a finger along your brow to simulate a large brow ridge and see if it makes much difference in protection, and whether this advantage would somehow translate into differences in survival or reproduction. Some have argued that large brow ridges also serve to keep the hair out of one's eyes. In fact, one anthropologist went so far as to fashion a set of artificial brow ridges that he wore to show that they did indeed keep his long hair out of his eyes![52] Although keeping hair out of the eyes might be an advantage of larger brow ridges, it does not seem likely that larger brow ridges would have actually evolved for this reason. If hair in the eyes did somehow impinge on someone's ability to survive, I would imagine that long hair could be more easily dealt with by cutting it off with a piece of sharp flint, or tying it back with a piece of hide. In addition, the hair hypothesis runs into problems when trying to figure out why brow ridge size reduced over time since *Homo erectus*—did long hair hanging in your eyes no longer matter? Another problem with hypotheses regarding the functional value of brow ridges is that they often treat the brow ridge as a single item being acted upon by natural selection and not part of the entire skull. We cannot analyze the evolution of individual parts of an anatomical whole without considering how they all fit together.

Several general models have been proposed to examine the anatomy of brow ridge size.[53] One is the biomechanical model, which looks at the effect of chewing stresses on brow ridge formation. Although no muscles are connected to the brow ridges, they are affected by stress generated in other parts of the skull. Use of the front teeth generates a certain level of stress, which may have been greater in some of our earlier ancestors who appeared to have used their front teeth in additional ways, such as a vise to hold objects (something we probably all still do against the advice of our dentists). The amount of force generated also depends on other features of cranial anatomy, including how much the lower face protrudes (i.e., is prognathic). The more prognathic an individual, the less efficient their bite, and the more force must be generated to achieve a given level of bite force. As bone responds to stress, this causes larger brow ridges. During human evolution, the size of the face reduced and our ancestors became less prognathic, changing the balance of forces and resulting in less stress from biting and, hence, smaller brow ridges.

Another model of brow ridge formation, known as the spatial model, looks at the relationship of the frontal bone (the part of the skull that contains the brow ridges) to the eye orbits and the brain case of the skull. The form of the frontal bone is influenced by the spatial relationship between these two parts and that in turn affects the development of the brow ridges. During infancy, the braincase overlaps the orbit and there is no brow ridge. As an individual grows up, the orbits advance and the separation of the face from the rest of the braincase leads to a brow ridge. The degree of separation is minor in some species and greater in others and is affected by overall size. During human evolution, the brain case increased in size and the face became less prognathic, and the interaction of these two cranial components led to a reduction in the average size of the brow ridges. It is worth noting that this model and the biomechanical model are not mutually exclusive, although debate continues about the relative influence of each.

As is often the case when examining anatomical evolution, the brow ridge is best understood as part of the total morphology. Finding interesting and subtle advantages of a trait, such as keeping the hair out of the eyes, and constructing elaborate evolutionary explanations ignores the morphological whole and reduces anatomical evolution to a hodgepodge collection of itemized traits. Brow ridge size reflects a complex set of developmental factors set within an evolutionary context and does not provide an index to a species' similarity to apes.

Myth #29 Neandertals walked bent over and were dumb brutes

Status: Early studies of Neandertals viewed them as primitive ape-like creatures that were quite different from modern humans. These once-popular images of Neandertals, including the ideas that they did not stand up straight and were dim-witted brutes, were due in part to inaccurate anatomical assessments and then-prevalent views of prehistoric humans. These negative images persist even today, and are clearly wrong. Although Neandertals were different in some ways from modern humans, they were fully bipedal, intelligent toolmakers and hunters.

Of all of our ancestors, Neandertals have probably received the worst press. To many, the very mention of Neandertals brings up images of brutish ape-men who walked bent over and lacked both intelligence and common sense. In short, they have often been considered prehistoric morons. The insults do not end there as the very word "Neandertal" has evolved into an insult. Typical dictionary definitions include "primitive, unenlightened, or reactionary; culturally or intellectually backward."[54] Why do Neandertals have such a poor reputation, and is it accurate?

The name Neandertal means "Neander valley" as the German for "valley" is "tal" and refers to a valley in Germany. A skeleton was discovered in 1856 at a cave site in the valley, and it, along with all similar fossils, have since been referred to as Neandertals. You may be familiar with an alternative, older spelling of "Neanderthal" with an "h" and perhaps have heard it pronounced with the "THAL" sound. Actually, the letter his silent regardless of the spelling. Since the middle of the nineteenth century, scientists have debated over whether Neandertals should be classified as a separate species, *Homo neanderthalensis*, or a subspecies of early *Homo sapiens*—*Homo sapiens neanderthalensis*. The debate centers on the interpretations of both the similarities and differences between Neandertals and us.

Neandertals are characterized by a set of traits that set them apart from earlier humans, such as *Homo heidelbergensis* and modern *Homo sapiens*. Neandertal fossils have been discovered in Europe, the Middle East, and some other parts of western and central Asia. Some Neandertal traits are present early in Europe and can be seen in *Homo heidelbergensis*, the likely ancestor of Neandertals.[55] Fossil specimens showing a full set of Neandertal features begin to appear between 100,000 and 200,000 years ago. The last Neandertals date to about 30,000 to 40,000 years ago in Europe, a time somewhat after anatomically modern humans had begun dispersing into Europe.[56]

Neandertal skulls show both similarities and differences when compared to *Homo heidelbergensis* and to modern humans. Brain size is very large in Neandertals. Although the average absolute brain size is larger than in many modern human samples, their body mass was also larger, as estimated statistically from long bone measures. As such, their brain size relative to body size is somewhat less than in modern *Homo sapiens*.[57] Neandertal skulls are long and low in profile and have large brow ridges that form distinct double arches above the eyes. Distinct features include the occipital bun, the large nose and inflated mid-facial region, and other cranial traits.[58] Some Neandertal features have been interpreted as long-term adaptations to a glacial environment, such as a larger nasal region to allow more effective warming and moistening of cold and dry air. Others have suggested that some cranial traits reflect the greater stress that Neandertals experienced by using their larger incisor teeth as tools. However, detailed statistical analysis of both Neandertal and modern human crania shows that another explanation for Neandertal uniqueness is genetic drift, and their cranial anatomy may simply reflect random shifts over time in an isolated population.[59]

Compared with early modern humans, Neandertals were relatively short and stocky and very muscular. Based on statistical estimates, the average adult Neandertal male was about 169 centimeters (5 feet, 7 inches) tall and weighed about 78 kilograms (172 pounds), and the average adult female was about 160 centimeters (5 feet, 3 inches) tall and weighed 66 kilograms (146 pounds).[60] Some of the arm and leg bones are more curved than in modern humans and there are some slight differences in the pelvis, but there is no doubt whatsoever that the Neandertals were bipeds.

Where then did the image of a bent-over and semi-erect Neandertal come from? By the turn of the twentieth century, several complete or partial Neandertal skeletons had been discovered in France. Complete and partial skeletons are more common by the time of Neandertals in part because the Neandertals buried their dead, something that was not done by earlier hominins. One of these skeletons came from a cave site dated to about 50,000 years ago in the French commune of La Chapelle-aux-Saints and consisted of an adult male that showed signs of aging. He had lost many of his teeth before death and suffered from arthritis, suggesting he was old, and in fact is often referred to as the "Old Man" from La Chapelle. He was actually probably less than 40 years old, which was still pretty old in those days.

The skeleton of the Old Man was analyzed and described in 1911 by French paleontologist Marcellin Boule.[61] Boule's reconstruction of the skeleton is the major reason a myth has grown about Neandertals waking about bent over. The reconstruction of the vertebrae suggested that the Old Man had a straight spine, which is typical of apes but unlike the S-shaped spine in humans. (Look at a drawing of a human skeleton from the side to see the two areas of curving.) Instead of the Old Man from La Chapelle having his head on top of his spine, as is the case for humans, Boule's reconstruction had the Old Man's skull out in front, just as we see when a living ape stands on two legs. Of course, an ape cannot balance with that type of structure; when an ape stands on two legs, they have to bend their knees to maintain an upright posture. According to Boule, the Old Man had the ape-like condition and would have walked upright in the same manner as an ape; that is, with a bent-kneed posture. In addition, Boule also reconstructed the foot and proposed that toes were divergent, another ape-like feature.

Boule's reconstruction contributed to the view that Neandertals were not quite human, and was therefore more similar to apes than to modern humans. When we add in the then-current interpretations of large brow ridges, cranial shape, and other traits as ape-like (recounted in the previous myth), we see the emergence of the familiar, though inaccurate, rendition of the bestial nature of Neandertals. Boule's view on human evolution was that modern humans were already around at the time of the Neandertals and that modernity had been in place for a long time, so that the Neandertals could not be part of our ancestry. To Boule, Neandertals represented a primitive divergence from a much earlier time, allied more with apes than with humans. Boule's analysis and interpretation carried great weight for many years. For example, in the 1947 edition of his classic text, *Up From the Ape*, Earnest Hooton, a preeminent physical anthropologist, wrote about the Old Man from La Chapelle noting that "The gait may have been a shuffling, bent-knee walk, in which the legs were not completely extended upon the thighs. On the whole, Neanderthal man must have been a rather gorilla-like type."[62]

Although the ape-like interpretation of Neandertal skeletal anatomy persisted largely in many expressions of popular culture, not everyone in the anthropological community accepted Boule's reconstruction, questioning different aspects. In 1957, anatomists William Straus Jr and A.J.E. Cave examined the Old Man's skeleton and found a number of problems with Boule's reconstruction.[63] Their examination showed that the vertebrae were affected by osteoarthritis more than Boule had thought, and after considering this pathology there was no evidence to support Boule's claim that the Old Man's spine differed from the spine of a modern human. They noted that given his arthritis, the Old Man might have had stood and walked a bit bent over, but "if so, he has his counterparts in modern men similarly afflicted with spinal osteoarthritis,"[64] and should not be used as a model for a healthy Neandertal. As we have learned even more about Neandertal skeletons, we can put to rest the idea that Neandertals walked bent over.

Although there are some differences between the skeletons of Neandertals and modern humans, these differences do not change the fact that the Neandertals were fully bipedal. It is interesting that Straus and Cave's analysis actually served to introduce a new interpretation of Neandertals that was 180 degrees from that of Boule. To Straus and Cave, there was very little difference between Neandertals and modern humans as shown in their famous quote: "if he could be reincarnated and placed in a New York subway—provided that he were bathed, shaved, and dressed in modern clothing—it is doubtful whether he would attract any more attention than some of its other denizens."[65] The idea that Neandertals were virtually identical to modern humans is now considered an overstatement that does not take into account the clearer differences, on average, between Neandertals and modern humans. I'm not sure what reaction a Neandertal in modern dress would attract—I suspect that you would notice certain differences if you spent time staring at the person—but I do not think you would run out of the subway as you might if encountered with the type of bestial Neandertal that was once in vogue.

Although Straus and Cave argued that Boule did not recognize the arthritis in the Old Man's skeleton, Neandertal expert Eric Trinkaus has noted that Boule was indeed aware of the diseased condition of the La Chapelle skeleton and attempted to correct for the damaged condition of its vertebrae. Much of Boule's reconstruction was based on normal bones (including other Neandertals) and was likely influenced more by his preexisting idea that Neandertals were an ancient species with no evolutionary connection to *Homo sapiens*, and these preconceptions affected his interpretation.[66]

As with the myth of Neandertal posture, the myths regarding the brutish and ignorant nature of Neandertals have also been dispelled. Neandertal brains were large and available information from endocasts show that their brains were similar in organization. We also know that Neandertals were skilled toolmakers. The stone tool culture of the Neandertals is known as the Mousterian and many of the tools are made using the prepared-core method described earlier (Myth 24), a method that takes creativity and intelligence. The stone tool technology of the Neandertals was also being used by early modern humans, so it seems a bit problematic to treat Neandertals that differently. Like modern humans, the Neandertals buried their dead, which is at least in part a symbolic behavior not present in earlier hominins. There is also ample evidence that they were proficient hunters of big game, again much like early modern humans.[67] However, as noted in an earlier myth (Myth 27), Neandertals did most of their big-game hunting up close, without access to long-range hunting technology, and paid the price in terms of physical trauma.

It is true that over time modern humans developed even more sophisticated types of stone tools, developed a technology built on using bone as well as stone, and invented long-range hunting weapons such as the atlatl and bow and arrow. There are also clear differences between Neandertals and later modern humans in terms of symbolic behavior. Whereas modern humans in the past 30,000 to 40,000 years are found in association with many examples of art and ornamentation, these behaviors are much rarer in Neandertals, and even the few examples available are contested. The strongest case for art and ornamentation is from the Grotte du Renne site at Arcy-sur-Cure in France, where rings and pendants have been found in association with Neandertal fossils, suggesting that they made these objects or perhaps traded for them with modern humans in the region. More recent carbon-14 dating of the site shows a wide range of dates and various materials may have been mixed together.[68] If so, then there is less evidence for Neandertals making art and ornamentation.

Discussion continues about the behavioral differences between Neandertals and modern humans. Neandertals were skilled toolmakers, but did not develop the extensive toolkit seen in Upper Paleolithic modern humans. Neandertals were proficient big-game hunters, but here too there are differences in weaponry and perhaps strategy. Although Neandertals buried their dead, they did so in shallow graves without grave goods as found for modern humans. We can argue about the significance of these differences, particularly in light of whether they have any bearing on the extinction of the Neandertals, but it is also clear that there are many similarities in intellectual and technological ability when we compare Neandertals and modern humans to earlier hominins and to apes. The differences between them and us, whether subtle or profound, pale next to the differences between apes and us. Given this, we can see that the view prevalent in the early days of paleoanthropological research that the Neandertals were essentially ape-like beasts in biology, behavior, and temperament no longer applies.

Neandertals definitely could not speak

Status: Reconstructions of Neandertal skulls have suggested that they might not have been capable of fully modern speech. Further analysis has tempered this conclusion to some extent. Anatomical, behavioral, and genetic evidence suggests that Neandertals could speak, although the extent of their linguistic ability is still being studied.

Could Neandertals speak? As we have seen in the previous myth, there continues to be debate over exactly how similar Neandertals are to modern humans. Although early ideas portraying the Neandertals as very ape-like are not credible, there is still much discussion over their anatomical and behavioral similarity to modern humans. Some see the Neandertals as subtly different, whereas others see a larger gulf between them and us. One area that continues to be debated back and forth is the subject of the language abilities of the Neandertals. Were they capable of modern human speech with its distinct sounds that can be uttered at a very rapid pace or did they speak more slowly and less clearly? Questions like this quickly bring up the larger issue of language origins in human evolution. When did language begin to evolve and when did it reach the capabilities of modern humans? There are long-standing debates over whether the shift from ape-like abilities to those of modern humans took place gradually over time or very rapidly and recently.

The obvious problem about the origins of spoken language is that we cannot hear it. (As a twist on the old tree-in-the-woods question, if a Neandertal spoke in the woods, would anyone ever know?) Apart from science fiction tales of time travel, we cannot observe spoken language in our ancestors and have to rely on indirect evidence of spoken language.

Part of that evidence is anatomical, as modern humans have evolved changes in both our brain anatomy and vocal anatomy reflect our ability to use language.

Because we cannot look at the actual brains of our ancestors, we rely on clues about brain anatomy from the analysis of endocasts, as described in previous myths. Ralph Holloway and colleagues have conducted extensive analyses of Neandertal endocasts and their possible implications for behavior. Although it is not possible to demonstrate with certainty that Neandertals *did* use language, Holloway and colleagues did not find any primitive features indicating that they were incapable of modern language abilities. They note that they "see no reason why Neandertals were not fully capable of human speech."[69] They suggest three basic stages in the evolution of the human brain, and by the final stage hominins had evolved large reorganized brains with modern cerebral asymmetries and hemispheric specialization corresponding to complex behaviors, including language. They conclude that this final stage was reached by the time that both Neandertals and modern humans evolved and may have been mostly completed in *Homo heidelbergensis*.[70]

Another way to look at language and the brain is to examine the archaeological record for evidence of behavior that might indicate language ability. As noted in the previous myth, Neandertals were capable of making sophisticated stone tools, hunted large animals, and managed to survive an inhospitable glacial environment for many millennia. On the one hand, it seems difficult to imagine how a large-brained species could possess all of these abilities without having some linguistic skill, either through gestural languages, spoken languages, or through a combination of both. On the other hand, we have also seen that there is little evidence for art or other symbolic behaviors other than burying their dead. Does this absence mean that Neandertals had not evolved to deal with abstract symbols, the kind of thing we would expect from a proficient user of language, or does it mean that they had a different type of language ability?

Another way to look at the Neandertal potential for spoken language is to consider how vocal anatomy has changed during the course of human evolution. The most dramatic change has been in the position of the larynx, also known as the voice box, which makes sounds. In apes and other mammals, the larynx is positioned high in the throat, restricting the range of sounds that can be made. As infants, we also begin life with our larynx positioned high in the throat, but during growth it descends, which increases the space for making sounds. By around two years of age, our vocal anatomy has changed allowing us to make a wider range of sounds, both vowels and consonants, and to be able to speak rapidly and distinctly. An interesting byproduct of this change is that after the larynx descends we are no longer able to breathe and swallow at the same time, something we are able to do when we are infants and still have the ape-like condition of a higher larynx. Our ability to speak comes with a price; after two years of age, we are more likely to choke because of the changes in the position of the larynx. The biological cost of an increased probability of choking to death (another "scar of human evolution") has been compensated for during human evolution by the adaptive value of rapid and distinct vocalizations.

Even though we have different vocal anatomy than apes, this is not something that is easy to examine in the fossil record because the larynx and most of the rest of our vocal anatomy is made up of soft tissue that decomposes. However, there are indirect ways of examining the likely position of the larynx in human fossils even though the larynx does not preserve. Phillip Lieberman and Edmund Crelin used anatomical landmarks on Neandertal skulls to estimate the position of the larynx and to reconstruct the entire vocal anatomy and modeled the range of sounds that a Neandertal could make. They concluded that although Neandertals had greater linguistic ability than apes, they could not make the full range of sounds possible for living humans, particularly vowels. Therefore, Neandertals represented an intermediate stage in the evolution of language.[71] The ability to make vowels is what allows human languages to be so rapid compared to other forms of communication (e.g., Morse code).[72] Based on Lieberman and Crelin's reconstruction, it would appear that Neandertals were not as proficient in this as were modern humans. However, many others have questioned their reconstruction and performed simulations that show Neandertals *were* able to make vowel sounds and speak rapidly.[73]

Some have suggested that the amount of bending in the base of the cranium correlates with the position of the larynx, and can be used to reconstruct vocal ability. Apes have a relatively flat cranial base, whereas modern humans have a highly flexed cranial base. Early studies used the amount of flexing as a proxy for larynx position, suggesting that language abilities had evolved by the time of *Homo heidelbergensis*. However, the association with cranial base flexing has been questioned, and it may not tell us anything about the position of the larynx.[74]

Another suggested correlate is the size of the hypoglossal canal, an opening in the base of the skull through which a nerve passes that controls tongue movements. It was suggested that the hypoglossal canal increased during human evolution to accommodate larger nerves that would be needed for more motor control of the tongue required for the evolution of spoken language. Preliminary results suggested that modern hypoglossal canal size had been obtained by 400,000 years ago, indicating that *Homo heidelbergensis* and Neandertals had spoken language as well as modern humans. However, comparative studies now show that there is no correlation between the size of the hypoglossal nerve and the hypoglossal canal, and canal size cannot be used to make any inferences about language ability.[75]

One useful piece of anatomical evidence is the hyoid bone, also known as the lingual bone, which is in the neck and aides in swallowing and movement of the tongue. The hyoid attaches by muscles to parts of the throat including the larynx. The shape of the hyoid bone and markings for muscle attachment can provide information on the position of the larynx. Hyoid bones are rarely found in the fossil record—only five have been found prior to the origin of modern *Homo sapiens*—one *Australopithecus*, two from *Homo heidelbergensis*, and two from Neandertals. One of the Neandertal specimens comes from a skeleton dating 60,000 years ago in Kebara Cave, in Mount Carmel, Israel, and the other from the El Sidrón site in Spain dating to 43,000 years ago. The Neandertal hyoids are similar in size and shape to those in modern humans, and different from those found in African apes and in *Australopithecus*. The anatomy of the Neandertal hyoid reflects a larynx that has descended in the throat, arguing for vocal ability similar to modern humans. The two *Homo heidelbergensis* hyoid bones come from the Sima de los Huesos site in the Sierra de Atapuerca in Spain and date to about almost half a million years ago. They are also similar in size and shape to modern humans,[76] suggesting that spoken language evolved prior to both Neandertals and modern humans, both of which are descendants of *H. heidelbergensis*. However, not everyone agrees on the significance of hyoid bone anatomy for inferring speech and research and debate will continue.

Another clue about language origins comes from other anatomical evidence discovered at the Sima de los Huesos site in Spain. Ear bones have been found for five specimens of *Homo heidelbergensis*. Measurement data were used to analyze the acoustic properties of the middle and outer ears of these specimens, showing that they had a frequency range for hearing similar to that of living humans and different from that of apes. *H. heidelbergensis* was capable of high sensitivity in the 2 to 4 kilohertz range in which much spoken language is perceived.[77] This level of sensitivity is found in present-day humans but not in chimpanzees. Although this evidence does not show that *H. heidelbergensis* was capable of speech, it does show that they were capable of *hearing* sounds associated with spoken language. Although this ability could have evolved for other reasons, it is consistent with the other evidence that spoken language had emerged by the time of *H. heidelbergensis*. If so, then we would expect to see the same abilities in Neandertals, their descendants.

In addition to archaeological and fossil evidence, further insight is now available from genetic evidence, specifically the analysis of ancient DNA extracted from Neandertal fossils. Ancient DNA will be discussed more at length in later myths, but for the moment we concentrate on one particular gene known as *FOXP2*. This gene codes for *forkhead box protein P2*, which is needed for brain and lung development. In humans, mutations of this gene have been linked to speech impairment and linguistic ability, which suggests that the gene may have a role in speech and language acquisition. This does not mean that *FOXP2* is "the gene for language" because complex traits and behaviors are generally not caused by a single gene. However, the nature of impairment for certain mutations does suggest that *FOXP2* plays some role in language acquisition and speech. There are only two amino acid differences between living humans and chimpanzees, and these key changes may be linked to the evolution of human language.

Today, genetic technology allows the detection and amplification of DNA, the genetic code, from fossil remains. DNA sequences were obtained from two Neandertal bones from the El Sidrón site in Spain. In both cases, the Neandertals had the human, and not the chimp, form of the DNA.[78] This finding does not mean that we can say without reservation that the genetic changes for human language had taken place in Neandertals. There are likely many genetic factors affecting the origin and evolution of human language and this is the only one we know of at present. However, it is important to note that what we do have is consistent with Neandertals having modern human language ability.

The evolution of language is incredibly complex and we do not have a smoking gun that points unambiguously to proof that Neandertals spoke and used language the same way as do living humans. The claim that had initially been made based on cranial reconstructions that Neandertals could not speak (at least in the same manner as modern humans) has been challenged. It is true that some of the evidence pointing toward modern human language ability has also been challenged, but remember that there are multiple sources of evidence. Collectively, the archaeological, anatomical, and genetic evidence strongly suggests that Neandertals had spoken language as did their immediate ancestors, *Homo heidelbergensis*. However, as is often the case with analysis of the fossil records, we will have to continue accumulating evidence to be definite one way or the other.

Myth #31 Modern humans appeared first in Eurasia

Status: Until the middle of the twentieth century, a number of scientists advocated that modern humans had their origin in Eurasia, starting with a focus on Europe and shifting more toward Western and Central Asia. As the fossil record increased and more accurate methods of dating sites were developed, it became apparent that modern humans first appeared in sub-Saharan Africa, followed by dispersal of some populations out of Africa into Eurasia, and then ultimately to Australia and the New World.

Where did modern humans come from? In a broad sense, we can say that modern humans evolved from earlier archaic humans (*Homo heidelbergensis*). What did this change involve, and when and where did it take place? We refer to ourselves (*Homo sapiens*) as modern humans and sometimes qualify this label as "anatomically modern humans" to note that we are physically very similar to the first modern humans in an anatomical sense, but have changed quite a bit in terms of our culture and technology. In addition to having large brains, we living modern humans have a relatively short and high well-rounded skull. We tend to have smaller brow ridges than our archaic ancestors. Out faces have receded, giving us a very flat face. Although our jaws and face are smaller, the lower jaw is marked by a prominent and protruding chin, something rarely seen in earlier humans.

It will probably be no surprise that skulls of humans from several thousand years ago show the same set of characteristics, as do skulls from 10,000 years ago. When did these traits first appear? Modern humans are found all over the world in the most recent past (within 20,000 years). Where did they come from, and when did they evolve from earlier humans? This may seem like a trivial question because in principle, the answer is simple—examine the fossil record across the world and note the earliest appearance of modern humans in each geographic region to find out where modern humans appeared first. In reality, the process is more involved. For one thing, it is not always easy to assign fossils unambiguously to one species or another given the range of variation that exists within species. Another problem is that we may not have complete coverage of all time periods for all regions, making the process of identifying the first modern humans analogous to completing a jigsaw puzzle with many pieces missing.

In the nineteenth century, when scientists began contemplating the implications of Darwin's ideas, there were only a few examples of ancient anatomically modern human fossils. All of these early examples were from Europe. Perhaps the most famous of these were the skeletal remains from the Cro-Magnon rock shelter in France, which we now date to about 30,000 years old. (This skull is shown in Figure 3.7 in Myth 24.) These remains have come to typify the ancient presence of modern humans in Europe. Indeed, the term "Cro-Magnon" has often been used as a label for early European modern humans as a group, and in some cases for all early modern humans throughout the world. By the end of the nineteenth century, numerous sites throughout Europe yielded remains of anatomically modern humans, which today we date to 35,000 or more years ago. Although modern humans were also found outside of Europe by the beginning of the twentieth century, they were not as old.[79] At the time, these dates suggested that the oldest modern humans were in Europe, possible evidence for a European origin.

However, during the late nineteenth and early twentieth centuries, many argued that Asia was a more likely place for the origin of modern humans. Some of these ideas go back to the views of German biologist Ernest Haeckel, who felt that humans more closely resembled Asian apes than African apes, and that therefore the birth of the human race was likely to have been in or near South Asia. He specifically suggested the mythical lost continent of Lemuria that some supposed existed in the Indian Ocean.[80] The notion of an Asian origin of modern humans was prevalent and gave rise to a number of expeditions to find ancient human remains. A Dutch physician, Eugene Dubois, moved to Indonesia in the 1890s in his search for human origins and did discover the first specimen of a more ancient species—*Homo erectus*. One of the most famous expeditions in search of Asian human origins was led by adventurer Roy Chapman Andrews, whose trip to the Gobi Desert in the 1920s uncovered the first known dinosaur eggs, although no human ancestors.

By the 1930s, attention turned to another part of Eurasia after excavations at Mount Carmel in Israel in Southwest Asia revealed several modern humans that were thought to be somewhat older than the European Cro-Magnons, and possibly their ancestor, and were often referred to as "proto-Cro-Magnons." Other nearby caves at Mount Carmel produced a number of Neandertal specimens, termed "Progressive" Neandertals because they did not have the extreme manifestation of many of the "Classic" European Neandertals. By the 1950s, a model had developed that intertwined the evolution of European and Southwest Asian Neandertals with that of modern humans. In 1957, paleoanthropologist F. Clark Howell proposed that early Neandertals in Europe underwent adaptation to a glacial environment resulting in the Classic Neandertals. He proposed that the early Neandertals in Southwest Asia underwent different changes and gave rise to the modern humans found at the sites of Skhul Cave at Mount Carmel and other moderns that had been found at Qafzeh Cave in the lower Galilee in Israel. The early modern humans from Skhul and Qafzeh later spread into Europe and the Neandertals died out.[81]

By the late 1950s, the Eurasian focus on the origin of modern humans had changed from eastern Asia to the Middle East and Europe, but Africa was still pretty much left out of things. By the 1960s and 1970s, there was of course increasing evidence of earlier hominin evolution in Africa, notably *Australopithecus*, *Homo habilis*, and *Homo erectus*, but no concrete evidence of early modern humans that were older than the modern humans that were found in Southwest Asia. This situation was changing by the 1980s when evidence began to accumulate that suggested greater antiquity of modern humans in Africa.[82] An alternative model was also developed at this time, which proposed that there was no single time and place for the origin of modern humans—the multiregional evolution model. In its initial form, multiregional evolution proposed that human populations had spread out across Africa and Eurasia at the time of *Homo erectus* and had remained connected by migration ever since, allowing populations across the Old World to share evolutionary trends. According to this view, different evolutionary changes had occurred in different places across the Old World, which were then shared throughout the species by migration and interbreeding. Modern humans arose from the coalescence of these changes.[83]

The fossil record has increased quite a bit since then, and we now have a much better idea of the geographic distribution of the earliest modern humans in different parts of the world. This evidence all points to a very early presence of anatomically modern humans in Africa, well before they appear elsewhere in the world. At present, the oldest definitive early modern human is a partial cranium from the Omo site in Ethiopia. Although this specimen was first discovered in 1967, accurate dating was not available until 2005, when it was discovered that it dated back 195,000 years ago.[84] Several other unmistakably *Homo sapiens* skulls were found in 1997 at the Herto Bouri site in Ethiopia that date back almost 160,000 years ago.[85] Several other sites in South Africa have provided fossils of early modern humans between 90,000 and 150,000 years old. Collectively, the fossil record shows that *Homo sapiens* was in Africa between 100,000 and 200,000 years ago.

By contrast, the earliest signs of anatomically modern humans elsewhere in the world are later in time.[86] The Skhul and Qafzeh sites, once thought to have been only about 35,000 years old, have been accurately dated to 90,000 to 100,000 years ago. The next oldest date for modern humans is from Australia, dating to at least 45,000 years, and perhaps a bit older. For East Asia, the dates are generally in the range of 30,000 to 40,000 years old. The oldest dates for modern humans in Europe range from about 30,000 to 40,000 years ago depending on the specific part of Europe. The Americas are not occupied until probably 15,000 to 20,000 years ago. We still lack a complete record of modern humans, but the evidence to date shows an early appearance of *Homo sapiens* in Africa, followed by the later dispersion throughout the rest of the world in the last 100,000 years. At this point, it seems very clear that modern humans first appeared in Africa, and not in any part of Eurasia.

A shift to an African origin is interesting in an historical sense. Africa had long been considered peripheral to the question of modern human origins and in the past several decades has proven to be the critical area with an African origin confirmed by the fossil evidence and supported by genetic evidence (more on this in the next myth). It is not surprising that it took some time for this view to develop as little work had been done for many years in Africa. Other factors include preexisting views (such as Haeckel's) supporting a Eurasian origin and examples of nationalistic pride in having what was viewed as the oldest human ancestor. (Witness the acceptance of Piltdown because it was found in England.)

Perhaps the most interesting aspect of an African origin of *Homo sapiens* is that when populations dispersed from Africa, they often moved into parts of the world where other humans were already living. The best example is Europe: Neandertals had already been living in Europe for quite some time before the arrival of modern humans and it looks like both Neandertals and modern humans overlapped in time in western Europe.

The coexistence of Neandertals and modern humans brings up a number of fascinating questions. We can ask if they interacted peacefully or not and if they traded artifacts (as some have suggested). However, from an evolutionary perspective, what we really want to know is if they interbred, and if they did, does any Neandertal ancestry remain in us today? We have ample evidence of both Neandertals and modern humans in the Middle East, living in the same place and using the same type of tools. There is evidence from ancient DNA (see Myth 33) that the two groups overlapped in time. There were no Neandertals in East Asia, but there were other human populations in the region at the time modern humans lived in Africa, such as fossils that some assign to *Homo heidelbergensis*, and perhaps even some late surviving groups of *Homo erectus*. The main point is that here too we have a case where modern humans moved into parts of the world where some form of earlier humans were already living, unlike the expansion into Australia and the New World, which had never been occupied by any early form of human.

Until recently, the question about the evolutionary relationship of modern humans to other humans has been addressed by two general types of models—the African replacement model and the assimilation model.[87] Both models agree that modern humans appeared first in Africa 200,000 years ago and some populations dispersed outside of Africa between 50,000 and 100,000 years ago. The difference between these models is whether the modern humans interbred with other humans outside of Africa. According to the African replacement model there was no interbreeding and the modern humans replaced earlier populations that became extinct. This replacement could have been due to a variety of causes, including the idea that modern humans were better adapted, biologically and/or culturally, and the other human populations, such as the Neandertals, became extinct because of competition. In strict form, this model states that modern humans were a genetically separate species incapable of producing fertile offspring even if they mated (much like donkeys and horses, which can interbreed and produce mules, but those mules are sterile, showing that the donkeys and horses are reproductively separate).

The assimilation model, however, states that some interbreeding occurred. Even though ancient human populations such as the Neandertals have gone extinct (no one has seen a Neandertal in over 30,000 years), they interbred with modern humans before they died out. In this model, the Neandertals (and perhaps other populations) were assimilated into the larger gene pool of modern humans dispersing out of Africa. Different variants of the model have proposed different levels of interbreeding with archaic humans, but all share the view that modern and archaic humans interbred to some extent. Some argue that the archaic humans such as the Neandertals are not really a different species, but perhaps a subspecies of *Homo sapiens*. Others point to the fact that many mammalian species have been observed to interbreed, such as the liger (a cross between a lion and a tiger). There has been considerable debate about which model provides a better fit to the fossil record and in the past few decades we have made use of genetic data, both from living humans and from fossil DNA, to answer this question, the subject of the next two myths.

Myth 32 "Mitochondrial Eve" is our only common female ancestor

Status: In the 1980s, genetic analysis of mitochondrial DNA, a form of DNA inherited only through the mother's line, demonstrated that a common female ancestor for all of humanity likely lived in Africa 200,000 years ago. Nicknamed "Mitochondrial Eve," this ancestor has sometime been interpreted incorrectly to be the only female ancestor alive at that time. She is the most common recent ancestor for mitochondrial DNA, but not for the entire genome, as different genes have different ancestral histories. Statistical estimates show that there were many other female ancestors alive at that time, but they did not contribute mitochondrial DNA.

The idea of a single male and single female ancestor is one that runs throughout Judeo-Christian culture. In Genesis, the human race begins with a single pair of humans created by God—Adam and Eve. In the late 1980s, genetic evidence relating to modern human origins was tied into this image with research on mitochondrial DNA, which demonstrated that all living humans are related to a common female ancestor. Because of the biblical imagery, this common ancestor soon became known as "Eve" in media reports, and received a technical qualification in other sources when she was named "Mitochondrial Eve." This qualification is necessary because she was our common ancestor for a small part of our genetic ancestry, our mitochondrial DNA.[88]

What is mitochondrial DNA? Most of our genetic code is located in the nucleus of our cells, arranged along long strands called chromosomes. Chromosomes come in pairs (humans have 23 pairs) and you receive one chromosome in each pair from your mother and one from your father. The genetic code is made up of four chemical bases: adenine, cytosine, guanine, and thymine (usually abbreviated as A, C, G, and T) whose arrangement controls the production and regulation of biological processes. Our 23 chromosome pairs contain a lot of genetic information—the total length of the human genome is over three billion bases in length! However, not all of our DNA resides in the chromosomes; a small amount (roughly 16,600 bases in length) is found in the mitochondria of our cells, outside of the nucleus. Most of us remember hearing about mitochondria in high school biology where we learned of their role in supplying energy to cells.

Although our mitochondrial DNA is but a very small fraction of our total genetic makeup, it has a very interesting and useful property—it is inherited only through the female line. When a sperm fertilizes an egg, the egg (from the mother) contains the mitochondria and the mitochondrial

DNA (abbreviated as mtDNA). The father does not pass along his mitochondrial DNA. This mode of inheritance is different from the DNA in our cells, which we inherit from both parents, who in turn received half of their DNA from each of their parents. Only females can pass their mother's mtDNA on to the next generation. If a woman has only sons, her mtDNA will not be passed on to the next generation. The inheritance of mtDNA makes it particularly useful for genetic studies of ancestry. For mtDNA, each of us has only one ancestor in any preceding generation. My mtDNA came from my mother, who got it from her mother, who got it from her mother, and so on. As a male, I did not pass on my mtDNA to my children.

An important focus of research on mtDNA and human origins is the concept of the "most recent common ancestor" (MRCA). You can figure out what the most recent common ancestor is for your mitochondrial DNA if you know how you are related to someone. For example, a person and their sibling both received their mtDNA from their mother, who is their most recent common ancestor. Likewise, a person and their first cousin on their mother's side have their maternal grandmother in common, who is their most recent common ancestor for mtDNA. Note that your common ancestry for mitochondrial DNA does not necessarily tell you anything about other ancestors. For example, I once attended a family reunion on my father's side along with a number of my first cousins. We all had two grandparents in common and shared a certain fraction of the DNA in our nucleus, but not our mitochondrial DNA. As these were my cousins on my father's side, I did not receive my mtDNA from the grandmother we had in common. Therefore, even though my first cousins and I had two ancestors in common (my paternal grandparents), we did not share the same *mitochondrial DNA* ancestor two generations back. Common ancestry in genetics is not necessarily the same thing as common ancestry in genealogical terms, and we do not have a single ancestry that covers all genes.

What makes mtDNA useful is the fact that the more time since two people have shared a mitochondrial ancestor, the more likely it is that mutations in mtDNA will have accumulated in one or both lines. When a mtDNA mutation occurs, it is then passed on through the female line. Over time, unique mutations occur that are then shared by all subsequent descendants, giving us genetic profiles (called haplotypes) that are combinations of different genetic forms. These profiles can reveal a lot about our mitochondrial ancestry, which in turn can tell us a bit about our history. In general, the more different two people's mtDNA sequences, the farther back in time their common ancestor existed. Because we have an idea about how fast mtDNA mutates, we can look at two mtDNA sequences and tell how far back in the past, approximately, they shared a common mitochondrial ancestor.

An important concept is that *any* pair of mtDNA sequences will have a common mitochondrial DNA ancestor at some point in the past. (Yes, any two of us, including you and me.) If we add a third sequence, there will be some point in the past when there was a single most recent common mitochondrial ancestor for all three sequences. The same applies if we add a fourth, a fifth, and so on. By extension, this means that *all* of humanity can trace their mitochondrial DNA back to a single mitochondrial DNA most recent common ancestor.

This basic principle forms the heart of what is known as coalescence analysis, but it can be initially a difficult concept to grasp. Figure 3.8 and Figure 3.9 provide an example of how this process works. Figure 3.8 is a genealogy covering five generations but only for mitochondrial DNA, which is why only women are shown. (Focusing on a trait inherited through one parent makes this process easy to visualize.) We imagine that there were five women in a population four generations ago, shown in Figure 3.8 as Abby, Adele, Alice, Amy, and Angela. Let us assume that each woman has a mate (not shown here because fathers do not contribute mtDNA) and each couple has two children. Each couple

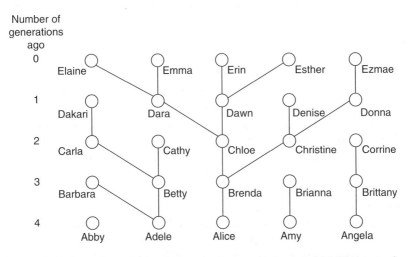

Figure 3.8 A genealogy showing the inheritance of mitochondrial DNA over five generations. Only females are shown as mitochondrial DNA can be passed on only through the female line. Some women have two sons, some have a son and a daughter, and some have two daughters. This figure was created choosing the outcome at random and only surviving females are shown.

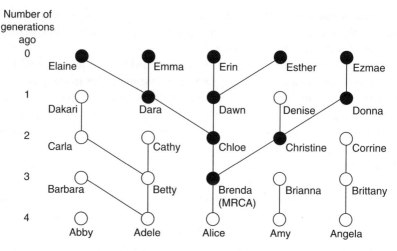

Figure 3.9 Identification of the most recent common ancestor (MRCA) of the mitochondrial DNA of five women (Elaine, Emma, Erin, Esther, and Ezmae) based on the genealogy in Figure 3.8. The black lines trace inheritance backwards until all the lines of all five women coalesce into a single female—Brenda—who is the most recent common ancestor for mitochondrial DNA.

could have two daughters, one son and one daughter, or two sons. All of the children will inherit their mother's mtDNA but only the daughters can pass it on to the next generation. In this hypothetical case, we see in Figure 3.8 that Abby did not have any daughters, so no mtDNA descendants are noted. However, Adele had two daughters—Barbara and Betty—and they both inherited Adele's mtDNA. We also see that Alice had one daughter (Brenda), Amy had one daughter (Brianna), and Angela had one daughter (Brittany). We continue to the next generation and see that Barbara and Brianna had no daughters, Brittany had one daughter, and both Betty and Brenda had two daughters each. This process continues up to the present day: Elaine, Emma, Erin, Esther, and Ezmae. By chance, some of the mtDNA lines have been lost.

Turning to Figure 3.9, we start at the present day and look backward in time, identifying common mitochondrial ancestors. To make this easier, the lines showing the mitochondrial DNA ancestry of the five women in the present generation are indicated by filled circles. We see that Elaine and Emma share a common mitochondrial ancestor one generation in the past; this ancestor is their mother, Dara (Elaine and Emma are sisters). Likewise, Erin and Esther are sisters, and their most recent common mitochondrial ancestor is their mother, Dawn. The fifth woman, Ezmae, received her mtDNA from her mother (Donna) and is not a sister of any of the other four women. If we go back another generation, we see that Elaine, Emma, Erin, and Esther all share the same mitochondrial ancestor, Chloe, who is their maternal grandmother. Go back an additional generation, and we see that all five women have a mitochondrial DNA ancestor in common—Brenda. What we see here is found in all genealogical structures—as we go farther and farther into the past, the lines of ancestry reduce and eventually coalesce to a single most recent common ancestor. For this analysis, Brenda is the most recent common ancestor for mitochondrial DNA.

What does this have to do with "Mitochondrial Eve?" In 1987, geneticist Rebecca Cann and colleagues published a study where they used mtDNA sequences from living humans to identify the likely time and place of the common mitochondrial DNA ancestor of humanity.[89] They examined the mtDNA of 147 women with maternal ancestry across the world, representing African, European, eastern and southeastern Asian, New Guinean, and Australian ancestry. They found 133 different mtDNA haplotypes. They then compared these haplotypes to each other to infer the patterns of genetic similarity that resulted from shared mutations. A family tree was constructed to show the genetic relationships of the mtDNA haplotypes, with more similar mtDNA haplotypes being located closer to each other on the tree. They found that the deepest branches in the genetic tree separated the sample into two groups; one group consisted of mtDNA haplotypes that came only from women of African ancestry, and the second group consisted of haplotypes from women of African and non-African ancestry. Cann and colleagues inferred that the only way this pattern could have emerged was if the most recent common mitochondrial ancestor lived in Africa, and then passed her mtDNA on to later generations in Africa, and then later to females that left Africa dispersing throughout the world. This was not completely unexpected, as it was known that *Homo erectus* had arisen in Africa and then dispersed to parts of Eurasia. The more controversial finding of their study was the estimate of *when* our common mitochondrial ancestor lived. Using estimates of the mutation rate for mtDNA, they concluded that this ancestor lived 200,000 years ago. To many, this date seemed too recent to be a genetic signature of the dispersion of *Homo erectus*, which had happened almost 2 million years ago. The mitochondrial DNA evidence, which has been confirmed by numerous studies, had been used by many to argue that the African replacement model was correct and there was no interbreeding with earlier human populations such as the Neandertals (a subject we come back to in the next myth).

From the beginning, the temptation to express these findings in terms of biblical analogy may have been irresistible. In the same issue of the scientific journal *Nature* in which the 1987 mitochondrial DNA study appeared, a commentary on the paper was titled "Human Evolution. Out of the Garden of Eden."[90] The Genesis imagery was also apparent in the January 11, 1988, cover of *Newsweek* announcing the title of the cover story in that issue—"The Search for Adam and Eve"—below which was an artist's depiction of a man and a woman (holding an apple) standing in front of a tree around which a snake was coiled.

Given these analogies and the constant reference to the common female ancestor as "Eve," it is no wonder that the idea that we had a single female ancestor in Africa 200,000 years ago took hold. However, the inference of a single female ancestor (and by extension, a single male ancestor) at one point in time is not correct when applied to the entire genome (all of our genetic material). Mitochondrial DNA will coalesce to a single individual, as shown in Figure 3.9, but she will only be a single ancestor for mitochondrial DNA. Other women could pass on the rest of their genome. For example, a woman who has only sons still passes on DNA to the next generation, just not her mitochondrial DNA. The misconception that we have a single common ancestor is misleading, as it applies only to mitochondrial DNA, which is why many qualify the original "Eve" nickname as "Mitochondrial Eve."[91]

The coalescent process works for other sections of DNA, allowing us to estimate the place and time of the most recent common ancestor for a number of genes and DNA sequences. Most of the coalescent analyses performed to date coalesce to an African ancestor but not all, and there is also variation in the date for these ancestors; some are relatively recent, such as mtDNA, and others are more ancient. The mathematics of coalescence analysis show that any DNA sequence will ultimately trace back to a single common ancestor, but not the same individual for each sequence. Our most recent common ancestors are spread out over time and space.

Another problem with the misconception of a single ancestor is the idea that this ancestor is the very first individual to exist. However, the most recent common ancestor also has ancestors. For example, in Figure 3.9, Brenda was the most recent common ancestor for mitochondrial DNA, but she had a mother (Alice) who was also an ancestor of all five women in the present generation. Likewise, "Mitochondrial Eve" also had a mother, who in turn had a mother, and so forth. The distinction here is that the common ancestor that can be identified using coalescence analysis is the *most recent ancestor*; we can track back genetically to this person, and no further, but it does not mean that this most recent common ancestor is an ultimate ancestor.

Although the research on "Mitochondrial Eve" is both fascinating and useful, it does *not* tell us that we had only one female ancestor for all genes in Africa 200,000 years ago. How many ancestors did we have? We can get an approximation of this number based on measures of genetic diversity. Because of genetic drift, smaller populations have less diversity than large populations. Making certain assumptions, we can get a rough estimate of average population size over time based on the level of genetic diversity we see in our species today. A number of studies have looked at this with a variety of genetic markers, and most suggest that there would have been about 10,000 individuals (male and female) on average during our species' lifetime. However, even this rough estimate can be misleading; it is an estimate of the number of reproductive individuals under a set of simplifying assumptions and does not necessarily mean that there were only 10,000 ancestors.[92] Depending on the demography of past populations, the actual number of ancestors might have been several times higher, and perhaps even a few hundred thousand.[93] Nevertheless, it definitely was not a single person or pair, except in the very restrictive sense of being our mitochondrial DNA ancestor.

Myth 33 Neandertals did not interbreed with modern humans

Status: Some have suggested that Neandertals were an extinct side branch of human evolution, a separate species that did not interbreed with modern humans and therefore are only our distant cousins. Ancient DNA analysis has produced a draft sequence of the Neandertal genome, which shows that between 1 and 4 percent of the ancestry of Eurasians comes from the Neandertals. Additional studies of ancient DNA has detected other archaic human groups that have also contributed to the ancestry of some living humans. The relationship between past and present populations is more complicated than once thought.

What happened to the Neandertals? As noted in an earlier myth, the last Neandertals lived in Europe 30,000 to 40,000 years ago. Their disappearance has fueled many speculations, both scientific and fictional (such as the novel, *The Inheritors*, by William Golding).

Explanations for the Neandertals' disappearance often involve the appearance of modern humans in western Europe, who replaced the Neandertals. Was this replacement a consequence of violence? Perhaps, although we see no evidence in the fossil record that Neandertals were deliberately killed off. Some models of Neandertal extinction suggest that they were a relatively small population to begin with, and were outcompeted by a larger group of incoming modern humans who possessed some advantage that allowed them to be more successful.

In any event, it is clear that the Neandertals are no longer with us as a distinct group. The larger question is whether they left any genetic material behind or not. Here, we return to the two views of modern human origins described in Myth 31—the African replacement model and the assimilation model. We want to know if there was any interbreeding of modern humans and Neandertals before the Neandertals disappeared as a distinct population. If there was some interbreeding, then some of our ancient ancestry is Neandertal, and Neandertal genes live on in us today. Under this view, the Neandertals were assimilated into a larger gene pool of modern humans, whose genetic contribution were much larger and swamped the genetic contribution of Neandertals.

Our relationship to Neandertals has been debated ever since the first Neandertal remains were discovered in the nineteenth century. Some have argued that Neandertals are similar enough (they are big-brained bipeds like us) to be closely related to us. The nature of this relationship ranged from the view that the entire sequence of human evolution went through a Neandertal stage to the view that they were a different subspecies of humans that, while different, never diverged too far from our line to be a distinct species and could interbreed with moderns. Others argued the opposite, citing numerous differences between modern humans and Neandertals, proposing that they were indeed a reproductively isolated species that became extinct and left no genes behind.

Since the nineteenth century, the debate has focused on anatomical similarities and differences (as well as insights from archaeology). Although there is consensus that the Neandertals are a physically different population from modern humans, the question of interbreeding has remained. Here, it has been difficult to translate anatomical differences and patterns of variation into any direct assessment of interbreeding. Some have noted patterns of regional continuity, where Neandertal traits persist, although at lower frequencies, in early modern humans in Europe before declining over time.[94] These patterns suggest that some interbreeding took place. Another indication of possible interbreeding is a 25,000-year-old skeleton from Portugal of a four-year-old modern human child who shows some

Neandertal characteristics, possibly reflecting recent Neandertal ancestry.[95] Although suggestive, there are alternative explanations for such evidence and it has not been conclusive one way or the other.

When I was in graduate school, we would joke that what we needed to solve the question of Neandertal interbreeding was the DNA of Neandertals. At the time (the late 1970s), such suggestions were not serious, as the idea of obtaining DNA from ancient remains was definitely in the realm of science fiction. Little did we know that advances in genetic technology, particularly the polymerase chain reaction, a way of amplifying small amounts of DNA, lay in the future. We now have Neandertal DNA, and although it does provide evidence of some interbreeding with modern humans, the resulting pattern is more complex than was once thought.

The first Neandertal DNA sequence was a small piece (only 379 base pairs in length) of mitochondrial DNA that was extracted from the original Neandertal fossil from Germany.[96] Although this sequence represented only a very tiny fraction of the mitochondrial DNA genome, it was large enough to provide some interesting comparisons with DNA from living humans. For example, there were 27 differences between the Neandertal and living human DNA sequences compared with only eight differences on average between any pair of living humans. The Neandertal DNA was clearly different, but it was not clear whether it was different enough to rule out interbreeding. Since that time, mitochondrial DNA sequences have been recovered from over 25 Neandertals, including several of the entire mitochondrial genome (over 16,000 bases in length).[97]

The mitochondrial DNA evidence did not support any interbreeding of Neandertals and modern humans.[98] Certain unique features of Neandertal mitochondrial DNA have never been found in any living human, which suggests that there was no interbreeding. However, one of the problems with mitochondrial DNA is that it is inherited as a single unit and is statistically equivalent to a single gene, which means we may only be seeing a fraction of evolutionary history. For example, it is possible that early modern humans interbred with Neandertals but their mitochondrial DNA was lost due to genetic drift (chance) over many millennia. This problem is analogous to the loss of a unique surname over time due to chance in family history, when a man has only daughters who will not pass on their maiden name.

Genetic technologies continued to develop and soon allowed a draft sequence of billions of bases of the nuclear DNA extracted from three Neandertal fossils from Croatia dating back 38,000 years ago.[99] The draft Neandertal sequence has provided many clues about the biological nature of Neandertals and their similarities and differences from living humans. Some of this information shows how selection may have been different for Neandertals and modern humans, including genes that affect metabolism and skeletal development.

In terms of evolutionary history, the Neandertal nuclear DNA sequence shows us that there was interbreeding between Neandertals and modern humans. Richard Green and colleagues used several methods to detect this Neandertal ancestry. For example, they compared the Neandertal DNA sequence with those of five living humans: two from sub-Saharan Africa, one from New Guinea, one from China, and one from France. We know that Neandertals and modern humans diverged from a common ancestor before modern humans left Africa. If these lines remained separate and there was no interbreeding with Neandertals, then the Neandertal DNA sequence should be equally different from all living humans today regardless of geographic origin, because we did not differentiate until *after* the split with the Neandertals. Instead, the comparisons showed that the Neandertal DNA sequence was more similar to Eurasian sequences and less similar to the African DNA sequences. Further comparisons allowed the researchers to determine that this similarity was due to gene flow from Neandertals into modern humans, and not the reverse.

The comparisons with Neandertal DNA allowed Green and colleagues to reconstruct a likely pattern of interbreeding in the past. Because the DNA evidence showed gene flow from Neandertals into people in Europe and Asia, but not Africa, this means that interbreeding with Eurasians took place after modern humans began dispersing out of Africa (and therefore that Neandertals did not interbreed with ancient populations in Africa). Further, the finding that both Europeans and Asians appear equally related to Neandertals meant that this interbreeding took place before Europeans and Asians split off from a common ancestor out of Africa. Given these constraints, it looks like this interbreeding took place in the Middle East. Under this model, Neandertals and the ancestors of modern humans diverged from each other several hundred thousand years ago. Modern humans arose in Africa by 200,000 years ago, and began dispersing out of Africa after 100,000 years or so, with some populations moving into the Middle East. Some limited interbreeding with Neandertals took place there before the modern humans dispersed into the rest of Eurasia, carrying Neandertal genes to all non-African populations. This model also fits the fossil record in that we have evidence that both Neandertals and modern humans lived in the Middle East at around the same time.[100]

The DNA sequence comparisons also allowed Green and colleagues to estimate the rate of Neandertal gene flow—between 1 and 4 percent of the nuclear DNA of living Eurasians comes from Neandertals. Although this rate of intermixture is low, it is not zero. Modern humans and Neandertals interbreeding may not have been an everyday event, but it did happen at least once, and shows that there was not a total replacement event. Neandertals are now extinct as a group, but some of their genes are still with us.

As we learn more about Neandertal DNA, it appears that the pattern of interbreeding may be more complex than at first thought. Further comparisons of the DNA sequences of Neandertals and living humans shows that East Asians are genetically closer to Neandertals than are Europeans, which means that the amount of Neandertal ancestry is likely to be somewhat higher in East Asian populations than in European populations. This result may seem surprising given that most Neandertal populations lived in Europe. There are two possible explanations for this pattern. First, there may have been additional interbreeding of Neandertals with the ancestors of living East Asians that took place after the divergence of European and East Asian populations. A second possibility is that there was an extended period of interbreeding in the Middle East, perhaps over thousands of years, and the ancestors of living Europeans diverged during this time, and had not accumulated as much Neandertal ancestry.[101] Further DNA sequences and analyses will help determine if either model is correct.

Another interesting finding is detection of a very small amount of Neandertal DNA detected in the Maasai population of East Africa. All other studies have shown no Neandertal ancestry in Africa, consistent with the model that interbreeding took place after modern humans left Africa. However, the Maasai results are actually more easily explained by recent gene flow into the population from outside of Africa, bringing in some Neandertal genes from outside of Africa. As the Maasai have a noticeable level of non-African ancestry, this is not surprising.[102] This finding shows us that any reconstruction of the past from living genetic data must deal with the fact that populations are always in contact with others. When we look at models proposing that human populations split from one another, such as groups leaving Africa or the divergence of Europeans and Asians, we do not mean that these groups will thereafter remain isolated from one another. Gene flow continues over time and a simple model of populations splitting, like branches of a tree, is a simplification.

The reconstruction of Neandertal interbreeding gets even more complicated due to the discovery of ancient ancestry from another, previously unknown population of archaic humans.[103] Excavations of Denisova Cave in Siberia have uncovered Upper Paleolithic tools and a few fragmentary fossils dating to between 30,000 and 50,000 years ago. The fossils found in the cave are by themselves not terribly interesting, consisting of some teeth, a finger bone, and a toe bone, all too fragmentary to assign to any particular species of *Homo*. The gold mine from Denisova Cave is the DNA that was extracted from the fossils, including mitochondrial and nuclear DNA sequences representative of an ancient group that we call the "Denisovans." The nuclear DNA of the Denisovans is different from both Neandertal and modern humans DNA, but more similar to the Neandertals, suggesting that the Neandertals and Denisovans shared a common ancestor after the split from modern humans. Based on the number of genetic differences, the divergence date of Neandertals and Denisovans is estimated to be 640,000 years ago.

Who were the Denisovans? This question is difficult to answer because we have the DNA but know very little about the anatomy of this population (and, as noted in the next myth, there are limits to what we can tell about human evolution without fossils). Could the Denisovans be the late surviving descendants of some other archaic group, such as *Homo heidelbergensis*, or were they some hitherto unknown archaic human group? Regardless of their identity, we have evidence that the Denisovans interbred with *some* ancestors of modern humans at some point during their existence; living Melanesians have between 4 and 6 percent of their ancestry from the Denisovans.[104] Additional studies have also detected Denisovan ancestry in Australian aborigines, Polynesians, and parts of Indonesia.[105] There are some interesting geographic questions here. Living humans with Denisovan ancestry are found in Southeast Asia and Australasia, but the fossils from Denisova Cave were from much further north in Siberia. It is likely that the Denisovans were once a geographically widespread group extending over much of eastern Asia, but only small pockets of admixture in the south have survived their extinction. There are also indications that the Denisovans in Siberia has some local Neandertal ancestry,[106] further showing a complex web of interconnecting populations.

It is interesting that some human populations today have both Denisovan and Neandertal ancestry. We are starting to see a complex pattern in archaic ancestry—some human populations today have Neandertal ancestry, some have Neandertal and Denisovan ancestry, and some have neither. Perhaps there are other archaic lineages awaiting discovery. Although many questions and mysteries remain about the genetic contribution of Neandertals and Denisovans to our ancestry, it is clear that the idea that Neandertals could not interbreed has to be rejected. Although most of our ancestry traces back to the first modern humans in Africa, there appears to have been a number of localized but significant interbreeding events that occurred when these early humans encountered preexisting archaic human populations.

We do not need fossils any more to learn about human evolution

Status: Recent analyses of ancient DNA of Neandertals and Denisovans provide new ways of examining the relationship of different groups during human evolution, giving detailed insights not available from the fossil record. However, the idea that paleoanthropological research in the future will be all or mostly based on ancient DNA, rather than fossils, is premature. Fossil evidence is critical to our understanding of human evolution, and paleoanthropology is a multidisciplinary science that must take into account data from fossils, archaeology, and ancient DNA, among other fields.

"Where do you dig?" I suspect that anyone who teaches biological anthropology gets this question quite often. When I reply that my research does not involve fieldwork excavating fossil remains, people sometimes seem disappointed. "I thought you said you studied human evolution?" is one rejoinder. I go on to explain that much of my research has been on very recent human evolution, including the genetic legacy of our species' origin and expansion out of Africa. This research does not involve fieldwork or even laboratory work, but computer analyses and mathematical modeling using publicly available data on cranial measures and genetic markers. People react to this in different ways (e.g., "That's interesting" or "Boring!") but in all cases I think the image brought forth is quite different than they expected, where I would be up to my neck in dirt or hunched over a table in a museum.

However, I have also noted that these responses have changed a bit in recent years and people are less surprised to find out that genetic studies have relevance for studying human evolution, either by looking at present-day patterns of genetic variation or ancient DNA. Thanks to magazines and television specials on genetics, ancestry, and evolution, the public is now aware that much of the research on human evolution now involves genetic analysis. The previous two myths have provided examples of how our understanding of modern human origins has been enhanced through the investigation of patterns of genetic variation in living humans and the analysis of Neandertal DNA.

The ancient DNA evidence, confirming our relationship to Neandertals as well as the presence of a previously unknown group (the Denisovans), demonstrates remarkable achievements that would have been considered fantasy even a short time ago. When we stop to consider the advances in molecular genetics as a whole as well as those in ancient DNA analysis, we might stop to wonder if we actually need to look at fossils anymore. Is our technology advancing to the point where we will be able to answer all of our questions about our ancestors by analyzing ancient DNA? Why do we need fossils at all, except of course as the source of the ancient DNA? Are the major issues of paleoanthropology best addressed solely through genetic analysis?

Although studies of ancient DNA have definitely provided new insights, and will likely continue to do so, I argue that they will complement but not replace studying the fossil record. There are limits to what we can do with ancient DNA (at least for the present). First, DNA is not easily preserved, particularly under certain environmental conditions, such as hot climates. Second, DNA samples can often become contaminated with DNA from other organisms, past and present. Third, DNA deteriorates over time, and is not likely to be useful for any organisms living back past a million years ago. (The sequencing of the genome of a fossil horse dating back between 560,000 and 780,000 years ago is close to the limit in terms of time depth at the time I write this.[107]) Although great advances have been made in ancient DNA analysis, there are still definite limits and only a small number of fossils can yield any sequences. Because of the limitations, we are unlikely to be able to sample any ancient DNA for some hominin species. Some of these problems may be solved in the near future, but such changes will not remove the need to look at the fossil record.

One of the most useful insights from ancient DNA is that it provides a precise means of looking at population relationships. As noted in the previous myth, knowledge of the Neandertal genome has allowed us to demonstrate some limited interbreeding with modern humans, an observation that was not as clear from the fossil record. However, because we are not likely to have DNA for all species and sites, we ultimately have to utilize the fossil evidence to examine, at least at a broad level, the relationships of hominins across time and space.

We also need fossils to tell us what our ancestors looked like. It is not enough to know their genetic makeup. Although we can make some statements about the appearance of our ancestors, such as discovering genetic evidence for light skin in Neandertals, we do not know the complete underlying genetic basis of most physical traits. Even if we had this information, it would only be part of the picture as the development of any organism depends on the expression of genetic information within a given environment and a particular life history. Only fossils can tell us about our ancestors' phenotype—what they actually looked like in a given time and environment.

Further, there are many aspects of an organism's life that cannot be determined from genetics alone, as we need to see the development of the phenotype during life. Comparative studies of fossil specimens give us insight into the growth and development of individuals in a species, helping us understand how similar and different species are from one another. For example, much work has been done concerning the growth patterns of Neandertals and how they differ from modern humans. One study showed that although both Neandertals and modern humans have large brains, the actual pattern of brain development after birth was different, as was cranial shape.[108] Another study shows differences in the pattern of dental development, where Neandertals show faster rates of dental maturation.[109] Another example comes from studies of cross-sectional geometry of arm and leg bones, including the area of different parts of the bone and its overall strength and rigidity. These studies examine the structure of bones from an engineering perspective, where the bone is analogous to a hollow beam subject to bending and other stresses.[110] Analysis of the long bones of ancestors such as the Neandertals has allowed insight into physical strength, movement and other activity, and mobility.[111] Such analyses, and their evolutionary implications, require fossils for us to understand *how* our ancestors lived.

Fossils also tell us about life events that are critical for understanding how a species lived, such as their diet and their diseases. The fossil record can provide information on when individuals died, which tell us something about their average longevity. For example, analyses of Neandertal teeth used to estimate age of death shows that few Neandertals survived into their thirties! From an evolutionary perspective, this means that living long enough to be a grandparent is something new in human evolution, having arisen with modern humans.[112] Another example of insights from fossils into aspects of life and death is the study of Neandertal injuries described in Myth 27. These (and other) aspects of the lives of Neandertals and other earlier hominin species can only be determined from fossils.

The study of paleoanthropology needs to incorporate evidence from *both* fossils and genetics. There has always been some tension between those who study genetics and those who focus on the fossil record.

We have seen an example of this in the debate over the timing of the split of African apes and hominins (Myth 10), where genetic analyses of living apes and humans suggested a much younger split than was once predicted from the fossil record. Although the genetic and fossil evidence was reconciled, there was still a tendency to want to choose a side in a debate over which source of data was better—genes or fossils? There was a similar debate for some time in the discussion over modern human origins.

As the field of paleoanthropology matured, it was realized that both fossils and genetics (including ancient DNA) provide insights that need to be taken into account. In addition, the study of paleoanthropology needs to take into account insights from other fields, including but not limited to studies of archaeology, primate behavior, paleoecology, geological dating, and many others. Each field brings in a different set of perspectives and methods that work best when integrated with each other. Elsewhere, I have stated, "*all* of these approaches are vital and necessary to truly understand human evolution. I see little point in arguing about which approach is "better" or in ignoring the potential contributions *all* types of data bring to paleoanthropology"[113] (italics in original). Ancient DNA analysis will continue to supplement, but not replace, fossils.

Myth #35 All recent human species had large brains

Status: A general pattern of human evolution has been an increase in brain size over the past 2 million years. By 200,000 years ago, the fossil record has been characterized by large-brained humans, including modern humans and archaic humans such as the Neandertals. It has seemed for some time that all recent humans had relatively large brains. This view has now been challenged with the discovery of a strange fossil nicknamed the "Hobbit," with a small body size and brain. Many consider that the Hobbit belonged to a dwarf species that branched off from early Homo *and that lived as recently as 50,000 to 60,000 years ago. The evolution of brain size may have been more complex than once thought.*

As discussed in Myth 22, the earliest species of the genus *Homo* had brains that were larger than the earliest hominins (such as *Australopithecus*) but still smaller than found in modern humans. That myth also showed how brain size has increased over the past 2 million years. By 200,000 or so years ago, the fossil record shows modern humans as well as the Neandertals, both of which were large-brained species. The fossil record suggested that recent human evolution involved only large-brained forms and that there were no small-brained species in the last few hundred thousand years.

This view has now been challenged based on some fascinating fossil hominin remains that were discovered in 2003 at Liang Bua cave on the island of Flores in Indonesia.[114] The most astonishing of these finds was a partial skeleton of an adult female given the specimen identification "LB1." One of her most noticeable traits was her height—she was no more than 106 centimeters (3.5 feet) tall! Such small stature is not unknown in present-day humans, and is found in a number of pygmy populations. However, the skull of LB1 was quite unusual, having a very small cranial capacity less than any specimen of *Homo* and more typical of *Australopithecus*—the most recent estimate is 426 cc.[115] This cranial capacity is much less than would be found in any present-day pygmy population. Further, the skull did not look like that of a modern human; it was lower and longer in profile than modern humans, the maximum cranial breadth was lower on the skull than in modern humans, and it lacked a chin.

The small size of LB1 soon earned it the nickname of the "Hobbit," the race of small humanoids found in the book and film versions of Tolkien's *Lord of the Rings* saga. The authors proposed that LB1 was not a small modern human, but instead was evidence of a new species—*Homo floresiensis*, the human from Flores. The naming of a new hominin species is always controversial and this one even more so because of its inclusion in the genus *Homo*, a group that has traditionally been characterized by larger brains.

Further discoveries and analyses have added to what we know about LB1 and her kin. Stone tools and animal bones have been found at the site. Additional skeletal remains from eight other individuals have been recovered from the site, although no other skulls to date. Initially, the remains were dated between 12,000 and 95,000 years ago, but newer analyses show that the range is a bit older—60,000 to 100,000 years ago for the fossils, and between 50,000 and 190,000 years ago for some of the tools thought to belong to this species.[116] As with the skull, the postcranial remains show a mix of primitive and modern characteristics.[117] The wrist bones are primitive and quite unlike modern humans, and the feet are quite large relative to body size.[118] Taken together, the consensus of many anthropologists is that LB1 represents a different species. Although this species is similar to *Australopithecus* in terms of brain size, it more closely resembles some early form of the genus *Homo*, such as *Homo erectus* or even *Homo habilis*.

The discovery of a new species is always exciting, but to find one with such a small brain size so recent in time is unprecedented and raises a number of questions. What is the explanation for this small brain size? How are they related to us and to earlier human species and how did they get to Flores? The discoverers of LB1 noted that it resembled *Homo erectus* in a number of cranial features, suggesting some sort of evolutionary link. Another clue is the location. Flores is not that far from other locations in Indonesia where *H. erectus* has been found, such as Java. Although close in overall distance, the island of Flores is actually more isolated from Java than you would think. Flores lies to the east of a geological feature known as the Wallace Line, which marks a region where islands have not been connected to continental landmasses during recent evolutionary time. During times of glaciation the sea level drops, and what are islands today in Indonesia were in the past connected to the southeast corner of the Eurasian landmass. This connection is how we think *H. erectus* got to Indonesia in the first place—they walked across the connecting land. However, this was not possible for Flores, because there is a deep ocean trench that keeps it isolated even when the sea level drops.

Although *H. erectus* could not have walked to Flores, we do know that some early hominin somehow got there because we find *Homo floresiensis* there. One possibility is that like other species, some were transported accidentally on floating islands, which are pieces of shoreline ripped out by storms that then wash up elsewhere. Although many will drown or starve before washing up on another shore, we do know that this process has been responsible for the movement of species across water in the past, such as the migration of monkeys from Africa to South America.

In any case, the similarity of LB1 with *Homo erectus* suggests that *Homo floresiensis* might be descended from *H. erectus* regardless of how they got there. The big question is what happened then and why they became so small. As noted in Myth 4, island dwarfism leads to reduced body size when there are limited food resources on an island. Many examples of island dwarfism are known among dinosaurs, birds, and mammals, including cases of dwarfed elephants (they are about the size of a cow), some of which lived on Flores. When *Homo floresiensis* was first announced, it was labeled as a dwarf species of *Homo erectus*. According to this model, some earlier populations of *H. erectus* became isolated on Flores and evolved into *H. floresiensis* due to island dwarfism. We do not know when this first started, although it is interesting to note that stone tools have been found on Flores dating back 840,000 years ago, evidence of earlier hominin presence (though no bones have been found for this early date).

The island dwarfing hypothesis has been questioned because the brain size of LB1 seems too small for even a dwarf species of *Homo erectus*. There is a regular relationship between brain size and body size in mammals. If we scale the average brain size of *H. erectus* down to the body size of LB1, then the expected brain size would be larger than what we actually observe. Because the brain size of LB1 is smaller than expected, some have suggested that it is not a dwarf species of *H. erectus*. However, other analyses have shown that the observed brain size of LB1 is not that far out of line with expectations of island dwarfism for *H. erectus*.[119] Another possibility is that *H. floresiensis* is descended from an earlier species of *Homo*, such as *Homo habilis*, which has a smaller brain size to begin with, requiring more realistic levels of dwarfing. However, we have no evidence for *H. habilis* ever having lived outside of Africa, and a link between it and *H. floresiensis* would require a more complex model than currently supported by available data.

In addition, comparative analyses of cranial shape place the LB1 skull closest to *Homo erectus* than to other hominin species, strengthening the link between *H. erectus* and *H. floresiensis*. We also have to consider variation with early *H. erectus*, as the specimen most similar in shape to LB1 is an early *H. erectus* skull from Dmanisi (a site in the country of Georgia), which also has a smaller brain size than later specimens of *H. erectus*.[120] We may find that the line leading to *H. floresiensis* derived from the earliest dispersal of earlier and more primitive examples of *H. erectus*.

Although there is growing consensus that *Homo floresiensis* is a distinct species that lived alongside *Homo sapiens*, not everyone agrees. Some anthropologists have argued that LB1 is a modern human whose small brain and other physical traits are the result of a disease. If true, then the features of LB1 that are seen by many as indicative of species status are actually the result of disease. One possibility that has been raised is that LB1 is a pygmy that suffered from microcephaly, a developmental disorder that results in a smaller brain. Analysis of a virtual endocast constructed from CT scans shows that LB1 has a number of derived features typical of hominins and does not show the characteristics found in microcephalic skulls. The shape of the LB1 skull is also different from microcephalics and more closely resembles *Homo erectus*. Other diseases have also been suggested to account for the small brain size of LB1, including Laron's Syndrome, a genetic disease affecting growth hormones, and endemic cretinism, resulting from impaired thyroid function. Analysis of cranial shape shows that LB1 is different from the skulls of modern humans afflicted with Laron's Syndrome or endemic cretinism.

In addition, many of the associated pathologies have not been found in LB1, or their suggested association has been questioned.[121] A later study claimed that LB1 might be a modern human with Down syndrome,[122] but this conclusion has also been questioned.[123]

The pathology argument has been weakened considerably, but has not been uniformly rejected yet. Additional crania would solve the issue. If other specimens had a similar small brain size and cranial shape, this would be conclusive evidence for species status as it would be statistically unlikely to find two or more skulls with the same pathology. However, we do get some insight by comparing the wrist bones of LB1 with those from at least one other individual from the site. The other wrist bones show the same set of features as in LB1, indicating that these are characteristics of a different species, and not pathologies of modern humans.[124] More recently, the new dating of the fossils to 60,000 to 100,000 years ago provides further support that LB1 and her kind were not diseased modern humans, as modern humans did not arrive in that area until later in time (about 50,000 years ago).[125]

Given that LB1 and the other fossils from Liang Bua cave likely belonged to a different species from *Homo sapiens* raises interesting question about brain size and behavior. The fossils have been found in association with fire, complex stone tools, and evidence of hunting, which suggest advanced cognitive abilities. If these activities were performed by *Homo floresiensis*, then what does this tell us about the importance of brain size? Traditionally, larger brain size (>600 cc) has been used as a defining characteristic of the genus *Homo*. However, we always have to remember to consider both brain size and structural reorganization. Although it is small, the virtual endocast of LB1 indicates structural reorganization that is consistent with such behaviors. As noted in an earlier myth (Myth 22), both brain size and neurological reorganization occurred during human evolution and there may be different combinations of these two changes that have occurred.[126] We need to know more about how island dwarfism affects brain size and structure to explore these issues further.

Another intriguing question concerns the fate of *Homo floresiensis*. The revised dates announced in 2016 indicate the youngest specimens lived about 60,000 years ago, and stone tools associated with the species have been found as recently as 50,000 years ago, the same time that modern humans arrived in the area. Although it might be a coincidence that the youngest evidence of *Homo floresiensis* and the oldest evidence of modern humans both occur at about 50,000 years ago, another possibility is that the two events are related. Did modern humans contribute to the extinction of *Homo floresiensis*, perhaps related to competition over limited food resources?[127] Further analyses might shed some light on this and other questions of the life and death of the Hobbit.

Notes

1. See Wood (2012).
2. Spoor *et al.* (2007).
3. Lieberman (2001).
4. Schoenemann (2013).
5. Johanson and Edgar (2006), Cartmill and Smith (2009), Schrenk (2013).
6. Villmoare *et al.* (2015).
7. Berger *et al.* (2015).
8. These data are from Schoenemann (2013). Only specimens with an estimate for both geologic age and cranial capacity where used. Specimens that were infants or immature, or had a range of cranial capacity estimates, were excluded. *Homo floresiensis* (the Hobbit) is not included. The midpoint age was used when sites had a range of geologic dates.
9. Holloway *et al.* (2004), Schoenemann (2013).
10. Falk (1992), Schoenemann (2013).
11. Aiello and Wheeler (1995).
12. One of the best accounts of Goodall's early work on chimpanzee behavior is her book *In the Shadow of Man* (1971).
13. Pruetz and Bertolani (2007).
14. Whiten *et al.* (1999).
15. A good introduction to early studies of language acquisition in apes is Savage-Rumbaugh and Lewin (1994).
16. Survival dates for East Asia are from Shen *et al.* (2009); dates for southeastern Asia are from Indriati *et al.* (2011).
17. Harmand *et al.* (2015).
18. Lepre *et al.* (2011) provide evidence that Acheulean tools from Kenya date back 1.76 million years ago.
19. Wood (2014).
20. Klein (2009).
21. Leakey (1979).
22. Potts (1984).
23. Klein (2009).
24. Potts (1984: 347).
25. Roebroeks and Villa (2011).
26. Klein (2009).
27. Berna *et al.* (2012).

28 Goren-Inbar *et al.* (2004).
29 Boaz and Ciochon (2004).
30 Weiner *et al.* (1998).
31 Shimelmitz *et al.* (2014).
32 Roebroeks and Villa (2011).
33 Conroy and Pontzer (2012).
34 Schick and Toth (1993), Boaz and Ciochon (2004). See Shick and Toth for excellent photographs of scanning electron microscope views of marks left by carnivore teeth and stone tools.
35 Boaz and Ciochon (2004).
36 Campbell *et al.* (2006).
37 Bermúdez de Castro *et al.* (1999).
38 Díez *et al.* (1999).
39 Thieme (2000).
40 Stringer *et al.* (1998).
41 Berger and Trinkaus (1995).
42 Klein (2009).
43 Russell (1985).
44 Lahr (1996).
45 Brues (1977), Lahr (1996).
46 Russell (1985), Reader (2011).
47 Reader (2011: 76).
48 King (1864: 96).
49 King (1864: 96).
50 See Trinkaus and Shipman (1993) for a comprehensive account of nineteenth-century views on Neandertals, human evolution, and race.
51 See Russell (1985) for a discussion of different views on the evolution of large brow ridges as well as Stringer and McKie (1996) and Boaz and Ciochon (2004).
52 See Stringer and McKie (1996: 91) for photographs of the artificial brow ridges used by the late anthropologist Grover Krantz.
53 Ravosa (1988).
54 As an example of the dictionary definition of Neandertal, see dictionary.com.
55 Arsuaga *et al.* (2014).
56 Harvati-Papatheodorou (2013).
57 Ruff *et al.* (1997).
58 Harvati-Papatheodorou (2013).
59 Weaver *et al.* (2007).
60 Harvati-Papatheodorou (2013).
61 See Trinkaus and Shipman (1992), Tattersall (2009), and Reader (2011) for further description of Boule and his reconstruction of the La Chapelle Neandertal skeleton. Trinkaus and Shipman is particularly detailed.
62 Hooton (1947: 328–329).
63 Straus and Cave (1957).
64 Straus and Cave (1957: 359).

65 Straus and Cave (1957: 359).
66 Trinkaus (1985), Trinkaus and Shipman (1993).
67 Harvati-Papatheodorou (2013).
68 Higham *et al.* (2010).
69 Holloway *et al.* (2004: 289.
70 Holloway *et al.* (2004).
71 Lieberman and Crelin (1971).
72 Lieberman *et al.* (1972).
73 See Cartmill and Smith (2009: 394–395) for discussion of the reconstructions of Neandertal vocal anatomy.
74 Lieberman and McCarthy (1999).
75 DeGusta *et al.* (1999).
76 Martínez *et al.* (2008).
77 Martínez *et al.* (2004).
78 Krause *et al.* (2007).
79 Klein (2009).
80 Reader (2011).
81 Howell (1957). See the explanatory figure on page 342 for an excellent summary of his model of the evolution of modern humans.
82 See, for example, Bräuer (1984).
83 Wolpoff *et al.* (1984, 1994).
84 McDougall *et al.* (2005).
85 White *et al.* (2003).
86 Cartmill and Smith (2009), Klein (2009), Conroy and Pontzer (2012).
87 Relethford (2008).
88 Lewin (1987).
89 Cann *et al.* (1987).
90 Wainscoat (1987).
91 Lewin (1987).
92 Relethford (2008).
93 Eller *et al.* (2004).
94 Wolpoff (1999).
95 Duarte *et al.* (1999).
96 Krings *et al.* (1997).
97 Lalueza-Fox and Gilbert (2011).
98 Hodgson and Disotell (2008).
99 See Gibbons (2010) for a summary of this research and Green *et al.* (2010) for the original article, both in the same issue of the journal *Science*.
100 Hershkovitz *et al.* (2015).
101 Wall *et al.* (2013).
102 Wall *et al.* (2013).
103 See Bustamante and Henn (2010) for a summary of this research and Reich *et al.* (2010) for the original article, both in the same issue of the journal *Nature*.

104 Reich *et al.* (2010).
105 Reich *et al.* (2011).
106 Pennisi (2013).
107 Orlando *et al.* (2013).
108 Gunz *et al.* (2010).
109 Smith *et al.* (2010).
110 Ruff and Hayes (1983).
111 See Trinkaus *et al.* (1999) for an example.
112 Caspari (2011).
113 Relethford (2007).
114 The Hobbit find was announced and described by Brown *et al.* (2004). A popular account of this and other finds at the site is given by Morwood and Van Oosterzee (2007).
115 Brown *et al.* (2004), Kubo *et al.* (2013).
116 See Sutikna *et al.* (2016) for the newest dates for *Homo floresiensis*.
117 Jungers *et al.* (2009).
118 Aiello (2010), Orr *et al.* (2013).
119 Kubo *et al.* (2013).
120 Baab *et al.* (2013).
121 Falk *et al.* (2009), Aiello (2010), Baab *et al.* (2013).
122 Henneberg *et al.* (2014).
123 Westaway, Durband, and Collard (2015), Westaway, Durband, Groves, *et al.* (2015). The first of these references is an online popular article that includes online comments from those on both sides of the pathology argument.
124 Orr *et al.* (2013).
125 Gramling (2016).
126 Falk *et al.* (2009).
127 Hoffman (2016).

4 RECENT AND FUTURE HUMAN EVOLUTION

It is tempting to consider the emergence of ourselves (modern humans), discussed at the end of Chapter 3, as marking the end of human evolution. However, we have continued to evolve as a species, both biologically and culturally as we have expanded across the planet and have developed new ways of adapting, such as agriculture. This section of the book examines a number of myths and misconceptions about our recent evolution, focusing primarily on the last 12,000 years. This section also examines a number of myths and misconceptions about human variation, a topic that is best understood in terms of past evolutionary events. Finally, I conclude with several myths looking at misconceptions regarding the future of human evolution.

Each of us has billions of distinct ancestors

Status: Because every person has two parents, the number of our distinct ancestors appears to increase exponentially each generation in the past, giving unrealistic estimates of billions of ancestors within the past two thousand years, more people than lived at that time. This discrepancy is explained by past inbreeding, which reduces the number of potential ancestors in the past. Models of common ancestry show that living humans share a common ancestor within the past few thousand years, and a few thousand years before that, every person then alive either has no descendants living now or is an ancestor of everyone on the planet.

---

50 Great Myths of Human Evolution: Understanding Misconceptions about Our Origins, First Edition. John H. Relethford.
© 2017 John Wiley & Sons, Inc. Published 2017 by John Wiley & Sons, Inc.

The study of human evolution is a very personal matter because these are *our* ancestors, connected to us across millennia. One can imagine (but not attain) a perfect genealogical record stretching back in time, generation to generation, to early *Homo sapiens* or even farther back. When we put our own existence in an evolutionary tree, we run into an interesting paradox regarding the number of our ancestors. Your most recent ancestors are your biological parents. Each of them had two parents, who are your biological grandparents. Of course, each of your four grandparents had two parents, which make up your eight great grandparents. As we look back farther and farther in time, the number of ancestors doubles each generation, giving 16 great-great grandparents, 32 great-great-great grandparents, 64 great-great-great-great grandparents, and so on.

We can show this doubling with a very simple formula: the number of potential ancestors is 2^t, where t is the number of generations in the past. For example, 10 generations in the past, the number of potential ancestors is $2^{10} = 1,024$, and 20 generations in the past, the number of potential ancestors is $2^{20} = 1,048,576$, a bit over a million. For this last number, if we assume about 30 years per generation,[1] you had over a million ancestors only 600 years ago (a little before Columbus went on his historic voyage). This number is big, but not incomprehensible (unless you were trying to list them all). If we go back 30 generations (900 years) we get over a billion (1,073,741,824) ancestors. This is over four times the number of people that were actually alive at that time! (Historical estimates suggest around 250 million people in the year 1100.[2])

The numbers get even more unrealistic further back in time. As we track into the past, the human population gets smaller, but the number of potential ancestors gets larger. By 40 generations ago (1,200 years ago), the number of potential ancestors is almost 1,100 billion, which is not only over thousands of times larger than the number of people alive back then, but larger than the estimated number of hominins that have ever lived![3]

It is clear that something is wrong with our reasoning because we could not have had so many potential ancestors even a short time ago in history. The key to understanding this paradox is the qualifying term "potential," which refers to the maximum number of ancestors under doubling and not the actual number. We have far fewer actual ancestors because of inbreeding. As a simple example, consider that the expected number of ancestors someone has four generations ago is $2^4 = 16$ great-great grandparents. However, suppose that this person's parents are second cousins, which means they share a set of great grandparents. Because there are two shared ancestors in both parent's lines, this reduces the number of distinct ancestors to 14 from the maximum of 16.

Although we often think of inbreeding only in terms of mating of close relatives, such as first cousins or closer, everyone is inbred if you go back far enough, which reduces the number of actual ancestors. Mating between fifth cousins, for example, does not lead to any genetic hazard, but does reduce the number of unique ancestors. Inbreeding also accumulates, so that a distant common ancestor is also likely to be inbred, and someone might be related through multiple lines; for example, a person's parents could be third cousins, fifth cousins, and ninth cousins all at the same time.

After a certain point in the past, we actually have fewer and fewer distinct ancestors because of this accumulated inbreeding, something referred to as "pedigree collapse." When we start counting backward in time, the number of our ancestors increases from two parents to four grandparents to eight great grandparents, and so on. Unless we have a number of close cousin marriages in our immediate past, the actual number of ancestors is close to the maximum expected under doubling each generation, and continues to increase. However, as we go further into the past, we increasingly see the effects of more distant cousin marriages accumulating and reducing the number of distinct ancestors. Thus, the number of our distinct ancestors increases back in time up to a point, and then decreases.

None of us has billions of *unique* ancestors. A consequence of having a smaller number of actual ancestors is that it does not take too long looking back in time to find a common relative between two people. The media is always fascinated to find genealogical connections between famous people. One example is the fact that President Barack Obama and President George W. Bush are 10th cousins once removed. ("Removed" means that the person is in a different generation; for example, your first cousin's child is your first cousin once removed.) In this case, the common ancestor was a native of Cape Cod who died in the mid-1600s. Other examples include Barack Obama and Brad Pitt, who are ninth cousins, and Hillary Clinton and Angelina Jolie, who are ninth cousins, twice removed.[4] We should not be that surprised at these findings as they are not unusual. Given adequate genealogical records, we each could show a connection between these same individuals and ourselves at some point in the past.

Most people do not have an exact record of all of their ancestors, but we can still get an idea about relatedness of people using mathematical models. Computer scientist Douglas Rohde and colleagues examined a number of models of ancestry and found some fascinating results about our common ancestry (all of us).[5] If we were able to trace back the genealogies of all living people, we would find common ancestors in the past.

Even though we do not have such genealogies, we know common ancestry is the case based on the application of probability theory to the counting of ancestors. Rohde and colleagues developed probability based models to estimate the date for our most recent common ancestor. Note that even though we have used the term "most recent common ancestor" when talking about mitochondrial DNA ancestry (Myth 32), its use here is different, as we are talking about the most recent ancestor on a family tree, who may or may not have contributed any particular genes to any given person today.

We have to remember that genetic and genealogical ancestry is not always the same. Consider, for example, your genetic ancestry from your most recent ancestors, which are your parents. You receive 50 percent of your nuclear DNA from each parent. Each parent received 50 percent of their nuclear DNA from their parents, your grandparents. You might therefore assume that you received 25 percent of your nuclear DNA from each grandparent. This is correct as a statistical average, but in reality, you might receive a little bit more or less from any given grandparent because of chance. For example, there is a 50 : 50 chance that your father will pass on a particular chromosome that he received from his mother (or father), thus not a certainty. To complicate things further, the chromosomal material is not passed on intact; sections of DNA recombine during the production of sex cells, meaning that your father could pass along a chromosome that is partly from both of his parents. The chance deviations will be even larger further back in the past, so that the amount of genetic ancestry you receive from each ancestor in any given generation will not be the same. This means that even though you had thousands of ancestors in the past, many of them may not have contributed anything to you genetically. Your genealogical ancestor (such as your great-great-great-great-great-great-grandparent) may not be a genetic ancestor. You may have a famous person in your family tree, but this does not mean you received any genes from them. For this reason, I draw a distinction between a most recent common *genetic* ancestor and a most recent common *genealogical* ancestor.

Rohde and colleagues have estimated that the most recent common genealogical ancestor of all of humanity lived only 76 generations ago, which is about 2,300 years ago, assuming 30 years per generation. This estimate is based on a fairly simple model where the world is divided into several regions connected by a small amount of migration. They also investigated complex models that incorporated the geographic distance between human populations across the world, likely routes and levels of migration, and variations in population density. Depending on the specific parameters, they estimated that the most recent common genealogical ancestor lived between 2,000 and 3,400 years ago.

This most recent common genealogical ancestor was a person who is the ancestor of every single person alive today. Remember that this ancestor is the most recent link between all living humans. As we go back further in time, we will find additional common genealogical ancestors. At any point in the past, some individuals will be these genealogical common ancestors, but others will have no living descendants because they or their descendants had no children. Only common ancestors show an unbroken line from the past to the present. Looking further back, we eventually reach a point in the past when every single person who was alive falls into one of two groups: either they had no descendants alive today or they are ancestor of every person in the world today. At this point in the past, everyone alive today has the same set of ancestors! This point in time is called the "identical ancestors point."

Rohde and colleagues found that the identical ancestors point is only a few thousand years earlier than the time for the most recent common genealogical ancestor. Their simple model gave an estimate of 5,000 years ago, and the more realistic and complex modes gave estimates between roughly 4,200 and 7,400 years ago. Splitting the difference suggests that approximately 6,000 years ago we all had the exact same set of ancestors. This is about the time that civilizations were starting to spring up from growing agricultural societies.

There is one caveat with the models developed by Rohde and colleagues—it does not apply to populations that until recently have been isolated from the rest of the world. Their case in point is Tasmania, an island off the shores of Australia that became isolated over 9,000 years ago and remained so until European contact in the nineteenth century. Because of that isolation the identical ancestors point would have to have been earlier than 9,000 years if Tasmanians were included. However, there has been European and Australian mixture into Tasmanians since the nineteenth century, and they too are now connected to a more recent identical ancestors point.

It may seem odd that every person on the planet today, regardless of their recent ancestry, has the same exact set of ancestors about 6,000 years ago. Indeed, this conclusion would seem to fly in the face of common sense, because we would expect people with the same ancestry to have the same physical appearance and genetic makeup, which is obviously not the case. You might wonder, for example, how a light-skinned European can have the same set of ancestors 6,000 years ago as a dark-skinned African. The answer lies in remembering the difference between genealogical ancestry and genetic ancestry. Although everyone has the same set of ancestors on a family tree, we have not all received the same proportions of our genetic ancestry from each. As noted earlier, you do not receive equal amounts of genetic ancestry from each ancestor in any given generation. Someone of recent European ancestry has obviously received more genetic ancestry from the common European ancestors than from other parts of the world. We are all related in a genealogical sense, but not necessarily to the same extent in a genetic sense. Nonetheless, we all do share recent ancestry and history to some degree.

Myth #37 The first Americans came from Europe or the Middle East

Status: Archaeological and geographic evidence has long favored a Northeast Asian origin of the first humans that moved into the Americas some 15,000 to 20,000 years ago. This view has been challenged in recent years by suggestions that Native Americans originated in Europe or the Middle East. Genetic data, particularly DNA markers, show this is not the case, and the point of origin for the first humans in the Americas was indeed Northeast Asia.

When Columbus set sail on his historic voyage in 1492, he already knew (as did many people) that the Earth was round; the disagreement at the time was over the circumference. Columbus's estimate was too low, and when he set off to find a shorter route to reach India and Southeast Asia and landed in the Bahamas, he thought he had reached the East Indies (the name then given to eastern Asia, and thus the reason the native inhabitants of the Americas were named "Indians"). It was soon realized that the Americas were a new land that was essentially unknown to Europeans, apart from some earlier limited exploration of the northeast coast of North America by the Vikings. After the initial contact of Columbus, Europeans pondered the origin of the indigenous peoples of the Americas. As was typical of the times, this question was investigated using biblical interpretation. One was that Native Americans were one of the Lost Tribes of Israel from the Bible.[6]

A connection between Native Americans and Asians was suggested in 1590, when Friar Joseph de Acosta noted that many animal species appeared to have moved from the Old World to the New World in the past, and humans should be no exception. Based on geography, the most likely route of migration would be between northeastern Asia and northwestern North America because that is where the two continents are closest, separated today by a narrow body of water known as the Bering Strait, which is about 50 miles wide.

In the twentieth century, scientists supported an Asian origin of the first Americans based on archaeology and biology and showed that these first migrants arrived here thousands of years before Columbus. By the latter half of the century, archaeologists had developed an explanatory model of how the first Americans moved from northeastern Asia. At certain times in the past when the Earth's temperature dropped, water became trapped in glacial ice, and the sea levels dropped, connecting areas of land. At several times, the drop in sea levels exposed a 1,000-mile-wide connection between Asia and North America known as Beringia (also known as the Bering Land Bridge, although 1,000 miles is a bit wide for a "bridge"). During these times, humans could disperse from Asia to North America in search of food. Large parts of North America were at times under glaciers, making further movement impossible. However, at different times in the past an ice-free corridor separated the two major glacial masses in North America. Human populations could move through this corridor and then travel down into the rest of North America and from there into Central and South America. This scenario is limited by the times of climatic change as the movement of humans had to occur while the land bridge existed and the ice-free corridor was open. In recent years, this classic model has been questioned, with some proposing movement along the western coast of North America rather than through the ice-free corridor and others suggesting the first arrival used boats rather than walk across Beringia.[7]

For many years, the entry to the New World for the Beringia model has been linked to what was thought to have been the earliest evidence of humans in the Americas, the Clovis culture. Named after a site in Clovis, New Mexico, the dates for the Clovis culture range from 13,600 to 13,000 years ago, and perhaps slightly younger. These people were able big-game hunters, frequently going after mammoths and mastodons. The major distinguishing archaeological feature of this culture is the Clovis point, a bifacial stone tool that has a characteristic shape with "fluting," the removal of a long flake from the base of the tool. Although long considered the first Americans, a number of sites have been found in the past few decades that predate Clovis and support a much earlier movement into the New World. Today, archaeological evidence supports human movement into Siberia by 40,000 years ago, and the first movement into the Americas most likely between 15,000 and 22,000 years ago.[8]

Regardless of continuing debate over dates of entry, as well as the question of how many migrations took place, archaeologists have reached a consensus that the first Americans were from Northeast Asia and point to both archaeological and genetic evidence supporting this conclusion. However, there has been some disagreement from both scholars and the public with occasional suggestions for an origin point elsewhere. For example, some archaeologists have suggested a European origin for the Clovis people based on the similarity of Clovis stone tools and methods of manufacture to those of the stone tools of the Solutrean culture, found mostly in France and dating between 16,500 and 22,000 years ago. According to proponents of this model, early Europeans could have traveled to the New World by moving along the edges of the ice that connected Europe and North America during the last major glaciation, adapting to a maritime diet and conditions.[9] In general, archaeologists have rejected this hypothesis, arguing that any similarities between Clovis and Solutrean tools are coincidental and do not reflect an historical connection.[10]

Archaeological questions about population origins can be answered in part by evidence from the genetics of living and ancient populations. The use of genetics to examine models of origin and population history is a major focus of the field of anthropological genetics, that is, the use of genetic data in an anthropological context to explore questions of population history and adaptation. In the case of Native Americans, both the genetics of living peoples and ancient DNA from prehistoric peoples confirm definitively that the first Americans came from Northeast Asia.

Biological similarities between East Asians and Native Americans have been observed for a long time. This similarity became even more noticeable from skeletal analysis of Native American remains, which shows traits common to both populations. An example is a dental variant known as a shovel-shaped incisor, where there is a ridge along the sides of the back of the front teeth. Shovel-shaped incisors are not that common in African Americans or European Americans (less than 15%), but are highly prevalent in East Asians (generally more than 90%). High frequencies are also found in Native Americans, suggesting an historical link with East Asian populations.[11]

Much stronger evidence for an Asian connection came during the latter half of the twentieth century with the accumulation of vast amounts of genetic markers from human populations around the world. In addition to different blood groups of interest to anthropologists (see Myth 3), methods were developed for detecting genetic variants in red blood cell proteins and enzymes as well as genetic variants in white blood cells.

When large numbers of markers had been collected, scientists were able to construct various measures of "genetic distance" that expressed the average genetic difference between pairs of populations. One of the most comprehensive studies using these traditional genetic markers was performed by Luca Cavalli-Sforza and colleagues, who examined 120 different genetic variants in 42 populations around the world. In several regional analyses, they compared nine different groups: sub-Saharan Africans, Europeans, Australasians, Pacific Islanders, Southeast Asians, Northeast Asians, Arctic Northeast Asians, Native Americans, and a composite group of Middle Easterners, North Africans, and South Asians. They consistently found that Native Americans were genetically the most similar to Northeast Asians and Arctic Northeastern Asians and less similar to Europe or other geographic regions.[12] These results make sense only given a Northeast Asian origin of the first Americans.

Definitive proof of an Asian connection has emerged with the rapid development of molecular genetic methods in the past several decades, particularly studies of mitochondrial DNA. As described in Myth 32, different combinations of genetic markers define mtDNA haplotypes. These haplotypes are identified by the presence or absence of certain DNA mutations in the DNA bases or deletions of certain sections of the DNA sequence. Different haplotypes can be placed together in a broader grouping known as a haplogroup, which is a set of haplotypes that share a common mutation. For example, assume a given mutation took place in the past, and was then passed on to subsequent generations. Some future lines might develop additional mutations, defining new haplotypes, but all of these haplotypes would still belong to the initial mutation that defines the haplogroup. In an evolutionary sense, all haplotypes within a given haplogroup share a common ancestor. There are also sub-haplogroups that allow greater precision in identifying evolutionary connections.

Over two dozen mitochondrial DNA haplogroups exist in living humans. Native Americans have one of five different mtDNA haplogroups known as haplogroups A, B, C, D, and X. The first four are found throughout the Americas, and haplogroup X has been found only in North America. These haplogroups are found in both living populations as well in ancient DNA extracted from skeletal remains. Because mitochondrial DNA mutates relatively fast, there is a fair amount of differences in haplogroups around the world. The most common haplogroups (A, B, C, and D) are only found outside of the Americas in East Asia. These characteristic Native American mtDNA haplogroups are not found anywhere else, a finding that again points to an East Asian origin of Native Americans.[13]

What about haplogroup X? It has been found at low frequencies in some North American native populations, but is relatively rare in the eastern parts of Asia. However, it has also been found at low frequencies in Europe, a finding that suggested to some the possibility that at least part of the ancestry of Native Americans came from Europe.[14] Is this genetic evidence that supports the Solutrean model? One problem with this idea is that there are actually different subtypes of haplogroup X, and the specific variant found in Native Americans is different from that found in Europe and Central Asia, indicating that there is no direct pattern of ancestry.

It is true that haplogroup X is very rare in Siberia, and has not been found in the populations is in the eastern part of Siberia, which is where we expect the ancestors of Native American to originate. However, if we take the absence of haplogroup X in eastern Siberia as a rejection of the long-accepted Asian origin of Native Americans, then we have to throw out every other piece of genetic and archaeological evidence that ties Siberia to the first Americans. There is a much simpler explanation for the absence of haplogroup X in eastern Siberia—genetic drift, which can lead to the extinction of an allele. Genetic drift leads to extinction very quickly when the initial allele frequency is low. Haplogroup X has a low frequency in parts of Siberia as well as North America, and it likely existed at low levels in eastern Siberia before it became extinct because of genetic drift.[15]

Nonetheless, the fact that haplogroup X has been found outside of eastern Asia has been used to support various mythical ideas of an origin for the first Americans somewhere else. Anthropological geneticist Deborah Bolnick and colleagues note a case where a claim was made that Native Americans came from the Middle East because haplogroup X is very common among the Israeli Druze population. This superficial similarity would again require that all of the other genetic evidence regarding the connection between eastern Asians and Native Americans be ignored. Further, the Druze population has none of the more common Native American haplogroups (no A, B, C, or D). Finally, the subtype of X found in Native Americans (X2a) is different from those found in the Druze and elsewhere in the Middle East (where they have subtypes X2*, X2b, x2e, and x2f). Subtype X2a is not found in the Middle East and the more distant relationship with other subtypes does not constitute a direct evolutionary connection between the Middle East and the Americas.[16]

Finally, further demonstration of the ancestral connection between East Asia and the Americas comes from analysis of the Y chromosome.

One of the 23 pairs of chromosomes inside the nucleus of human cells is the sex chromosomes. There are two types—X and Y—where women have two X chromosomes and males have an X chromosome and a Y chromosome. To be a male genetically, you inherit the Y chromosome from your father. Most of the Y chromosome does not recombine with the X chromosome in males, so that most of a father's Y chromosome is passed on to his son, and then to his son (if any). As with mitochondrial DNA, there are a number of Y chromosome DNA haplogroups around the world. All Native American males have Y chromosome haplogroups C and Q, which are also found (no surprise) in parts of East Asia.[17]

Further insights have recently been revealed from analysis of ancient DNA. The entire genome of a male infant from a Clovis burial site (over 12,500 years old) was sequenced. When compared with the DNA sequences of living humans, this child's DNA was most similar to that of living Native Americans and not to populations elsewhere in the world. To many geneticists, this finding rejects the hypothesis that the Clovis people came from western Europe. Instead, the child's DNA shows that he belonged to a population ancestral to many contemporary Native American populations.[18] This view has also been confirmed by the recovery of mitochondrial DNA from the skeleton of a young girl who lived between 12,000 and 13,000 years ago in a submerged cave in the Yucatan Peninsula of Mexico. Her mtDNA was a subgroup of haplotype D, confirming a genetic relationship with living Native Americans.[19] Ancient DNA has also been used to resolve the question of the ancestry of a fossil known as Kennewick Man, discovered in Washington State and dating back almost 9,000 years ago. Some early analyses had suggested a possible European ancestry and later work on cranial similarity suggested greater similarity to East Asian and Polynesian populations than to present-day Native Americans. The question of Kennewick Man's ancestry was resolved in 2015 when his DNA was sequenced, showing the greatest level of similarity to present-day Native Americans and not Europeans or Polynesians.[20]

Although the genetic, skeletal, and archaeological evidence rules out a European origin of Native Americans in terms of an Atlantic migration, there is some interesting evidence showing a genetic connection, although not as usually envisioned. Ancient DNA analysis has been performed on a 24,000-year-old skeleton from Siberia. The nuclear DNA shows similarity with living Native Americans, although the mitochondrial DNA and Y chromosome DNA shows affinity with populations in western Eurasia, including some in Europe. Further, the nuclear DNA does not show any close relationship with East Asians. These results suggest a widespread group in Eurasia may have contributed to the ancestry of human populations in both the western and eastern parts of the large Eurasian land mass, leading to some genetic connections between living Europeans and Native Americans. The lack of close relationship between this ancient skeleton and living East Asians may indicate that the skeleton belonged to a population ancestral to Native Americans after diverging from an East Asian ancestral group.[21] As always, population history in our species is more complex than first thought.

Myth #38 The first Polynesians came from South America

Status: Polynesian populations are spread across the Pacific Ocean. A once-popular idea was that the Polynesians originated from South America, and adventurer Thor Heyerdahl showed it was possible to sail from South America to Polynesia. However, genetic data show that the Polynesians arose in South Asia or Southeast Asia and sailed eastward to colonize Pacific islands. DNA markers show this expansion was accompanied by moderate admixture with other Pacific Islanders.

We have seen that modern humans spread to Australia and the Americas, new lands previously uninhabited by earlier human populations. Another major expansion of modern humans was the expansion into the Pacific Ocean several thousand years ago. Today, anthropologists generally recognize three groups of Pacific Island populations: Micronesia, Melanesia, and Polynesia. Micronesia, which means "small islands," is an area east of Asia, consisting of more than 2,000 small islands. Melanesia is a region north of Australia that includes Papua New Guinea and the Solomon Islands, among others. The word "Melanesia" translates as "dark islands," so named because of the very dark skin of the native populations. Melanesia was first occupied by migrants from Australia at least 35,000 years ago. The largest geographic region in the Pacific Islands is Polynesia, which means "many islands." The boundary of Polynesia forms a rough triangle across much of the Pacific Ocean, extending to Hawaii to the north, New Zealand to the south, and Easter Island to the east.

Where did the Polynesians come from? The overwhelming current consensus is that they came from South Asia or Southeast Asia thousands of years ago, spreading eastward over time. This view was not always accepted. In the 1940s, the adventurer and anthropologist Thor Heyerdahl argued that Polynesians originated in South America and sailed westward into the rest of Polynesia.[22] Heyerdahl was struck by the presence of stone statues and pyramid-like structures on different Polynesian islands, which he felt were similar to those found with prehistoric civilizations in South America. Heyerdahl also noted possible similarities in folklore, which he felt could be explained by a connection. Among the Incas of Peru, the legendary high priest Kon-Tiki and several followers were said to have fled from a battle in Lake Titicaca in Peru westward into the Pacific Ocean. Heyerdahl argues that Kon-Tiki is the same person cited by some eastern Polynesian groups as their founder. Heyerdahl concluded that Polynesia had been settled initially from South America, followed by dispersal from east to west across the Pacific Ocean. Heyerdahl dismissed possible origins in or near Australia, Indonesia, or Asia because he felt the inhabitants would have been too primitive.

Heyerdahl noted that the first Polynesians would have to have been capable of sea travel. His main question was whether technologically simple watercraft could make the journey from South America to eastern Polynesia. To test this idea, he constructed a balsa wood raft, named the *Kon-Tiki*, using simple methods and sailed successfully to Polynesia from Peru. Heyerdahl stated that this successful voyage only showed that the trip *could* have happened, but not that it *did* happen. Nonetheless, his voyage (chronicled in his book, *Kon-Tiki*) became known throughout the world and attracted public attention.

Archaeological and genetic evidence has shown since that Heyerdahl's South American origin hypothesis is incorrect, and the first Polynesians instead came from Asia and sailed from the west to the east, and not the other way around. Archaeological evidence shows Polynesian populations ventured into the Pacific from Asia within the last several thousand years, reaching Hawaii and Easter Island about 1,600 years ago and New Zealand about 700 years ago.

Although humans had long possessed boat travel, much of it had been voyages to close destinations within line-of-sight. The expansion of Polynesian peoples into the Pacific Ocean is much more difficult, relying on their skills both as navigators and as sea voyagers. Navigation was made possible by relying on natural clues, such as the patterns of migrating birds and star constellations, rather than elaborate equipment. The typical sea craft used even today in Polynesia is a canoe, but of a special sort known as an outrigger canoe. Here, a log is attached parallel to, but several feet away from, the main canoe. Some canoes are double-outriggers, which have a log on both sides. The advantage of outrigger canoes is that they maintain stability of the main part of the canoe because it minimizes rocking back and forth.[23]

Archaeologists have long maintained that the Polynesians originated in South Asia, specifically Taiwan, or perhaps somewhere in Southeast Asia. The first Polynesians are the descendants of what is known as the Lapita Culture, named after a characteristic type of pottery that features elaborate geometric designs that are easily identifiable in the archaeological record. Lapita pottery first appeared between 3,200 and 3,500 years ago in the Bismarck Archipelago near New Guinea and then spread eastward into Polynesia via outrigger canoes over the next several hundred years. The dates for islands with Lapita pottery (and other distinguishing artifacts) show a clear west to east movement out of Asia into the Pacific Ocean, opposite to that expected under the Heyerdahl hypothesis.[24]

Genetic evidence also shows the Asian origin of the Polynesians. If Heyerdahl was correct, then Polynesians would be more closely related to Native Americans, particularly those in South America, than to Asians or Southeast Asians. The opposite has actually been found. For example, a study using 120 different genetic markers of the blood shows that Polynesians (as well as Micronesians and Melanesians) cluster genetically with populations from South Asia and Southeast Asia, and are more distinct from Native Americans.[25] The same patterns are found with more recent analyses of DNA markers. The Polynesians are not of South American origin.

There are actually variants of the Asian origin model. When we track the movement of Lapita pottery out of South/Southeast Asia from west to east we see that the first Polynesians would have passed near parts of Melanesia. Because people had been living in parts of Melanesia for 35,000 years, this means that there was possible cultural and genetic contact between Polynesians and Melanesians as the Polynesians moved by. Two models of Polynesian dispersal have been developed that consider the possible contact between these groups. One model, aptly named the "express train" model, posits that the movement out of Asia was very fast, and the Polynesians sailed past Melanesia with little or no contact, much like an express train with no stops.[26] Here, the first Polynesians moved through Melanesia fast enough so that there was little if any genetic mixture with Melanesians. An alternative, the "slow boat" model, has the same Asian origin and west to east expansion, but argues that the dispersal was not so fast, and there was some significant genetic contact between Polynesians and Melanesians.[27]

Which model is correct? These models can be contrasted in terms of their genetic predictions. If the express train model is correct, then there should be no evidence of genetic admixture with Melanesians in living Polynesians, whereas the opposite should be apparent if the slow boat model is correct. The actual testing of these models gets complicated because of possible genetic admixture in recent times. Classic genetic markers of the blood do not provide much resolution here. For example, the analysis of 120 markers mentioned earlier does not adequately distinguish between Pacific Island populations to support either model.

More insight has been gained from DNA marker analysis that allows better resolution of patterns of migration and mixture. The mitochondrial DNA evidence clearly supports the express train model. For example, genetic analysis has revealed a mtDNA haplotype named the "Polynesian motif" that is a set of genetic changes that is very common in Polynesians. There is a clear spatial pattern for this haplotype; it occurs in relatively low frequency in Asia but increases steadily in Polynesian populations from west to east, tracking their migration.[28]

Although the mitochondrial DNA evidence supports the express train model, studies of Y chromosome DNA have shown greater genetic affinity with Melanesians as predicted from the slow boat model. For example, one Y chromosome haplotype was found in high frequencies on Cook Island in Polynesia, absent in South Asian populations, but found in Melanesia.[29] The disparity of results from mitochondrial and Y chromosome DNA may reflect in part the greater effect of genetic drift on these markers. Because these types of DNA are inherited through one parent, genetic drift plays a greater role. In addition, both mitochondrial and Y chromosome DNA represent only a fraction of possible genetic variation and population history.

Greater resolution has since been provided by comprehensive analyses of nuclear DNA markers. Anthropologist Jonathan Friedlaender and colleagues examined two types of nuclear DNA markers in a large number of Pacific Island populations.[30] One type of marker, used in many studies of anthropological genetics, is based on microsatellite DNA, which are small DNA sequences (typically 2 to 4 bases in length) that are repeated in tandem. Different DNA sequences show different numbers of repeated units. For example, consider two DNA sequences, where one is "CACACACACA" and the other is "CACACACACACA." Both sequences show the fragment CA repeated multiple times; the first sequence repeats six times and the second sequence repeats seven times. Differences in repeat numbers are due to mutation, which can increase or decrease the number of repeated units. The second type of DNA marker that Friedlaender and colleagues looked at was "indels," which are insertions and deletions of DNA bases. As with microsatellite DNA, indels are caused by mutation. Their study examined a large number of markers—687 microsatellites and 203 indels—providing an excellent picture of population relationships. They found that the two Polynesian populations they included in their study (as well as Micronesia) were most similar genetically to Taiwan aborigines and other Asian populations, and more distinct from Melanesian populations, suggesting greater genetic input from East Asia.

Manfred Kayser and colleagues also conducted a DNA marker analysis of Polynesian origins, using 377 microsatellite markers to examine the possibility of dual ancestry.[31] Using these data, they estimated the average ancestry in their Polynesian sample was 79 percent East Asian and 21 percent Melanesian. These results were more in line with the slow boat model as the level of Melanesian ancestry was greater than expected from the express train model. Kayser and colleagues note that the typical results from Y chromosome DNA, which suggest much more Melanesian admixture, is likely a reflection of sex differences in migration. Here, there may have been more frequent admixture with Melanesian men than with Melanesian women.

In sum, the archaeological and genetic evidence shows that Polynesians originated in South Asia or Southeast Asia (the origin point is still being debated) and spread quickly into the Pacific Ocean, moving eastward all the way to Easter Island. Although we are still looking at the fine points of the Polynesian expansion, it is clear that Heyerdahl's South American origin hypothesis has been rejected. However, the rejection of a South American *origin* does not mean that there was *no* contact between Polynesia and South America in later, pre-Columbian times (before European contact).

Scholars have long been puzzled by some evidence of contact between these parts of the world, with some pointing to similarity of certain words and tools, suggesting the possibility of some cultural contact, although many question the linguistic evidence. There is also evidence that the sweet potato, an indigenous South American crop, was introduced into parts of Polynesia in pre-Columbian times. If so, then how did it get there given the lack of human genetic evidence for a South American origin? In recent years, it has been suggested that Polynesians continued eastward past Easter Island to land in South America around the year 1200 and then returned to Polynesia bringing the sweet potato (and perhaps other animals, such as chickens) and some artifacts back with them.[32] If so, then the Polynesians arrived in the Americas a bit earlier than Europeans did.

Myth 39   The origin of agriculture led to an improvement in health

Status: The Agricultural Revolution was a major event in recent human evolution, leading to population growth and radical changes in the structure of human societies. The origin of agriculture has long been viewed as an advance in human progress, but is also associated with a number of health problems in earlier times, including a decrease in nutritional quality and an increase in epidemic infectious disease.

A key event in human evolution was the Agricultural Revolution. Until recently in human prehistory, human populations all subsisted by hunting and gathering, a lifestyle that limited the size of both local populations and the entire species. The shift to agriculture meant that many more people could be fed. The ability to domesticate plants and animals led to surpluses of food and labor, which in turn changed the nature of human societies, including specialization of labor and social stratification. Agricultural populations grew in size and complexity, eventually developing into state-level civilizations. Much of what we view as human intellectual and technological progress comes from these beginnings. However, the negative consequences of the shift to agriculture are sometimes glossed over. It has sometimes been assumed that the quality of human life increased with agriculture, but in reality, the origin of agriculture leads to what Jared Diamond refers to "mixed blessings."[33]

The beginnings of agriculture take place in the Stone Age, specifically the Neolithic, or New Stone Age. About 12,000 years ago, some human populations began domesticating plants and animals. Although some of the earliest evidence of agriculture comes from the Fertile Crescent in the Middle East and North Africa, this does not mean that agriculture arose but once and then spread out from there. Agriculture actually was independently invented in seven different core areas in Asia, Africa, and the New World. An exception to this pattern is in Europe, where agriculture diffused from the Middle East.[34]

The specific causes of the change to agriculture are likely to have varied from place to place, but in general appear to have involved some combination of environmental change (the climate was warming up, making agriculture possible) and population pressure (an increasing human population that was becoming difficult to feed relying on just hunting and gathering). The shift to agriculture was fast, but did not happen all at once; over time, human populations began relying more and more on agriculture. Fertility rates increased with the origin of agriculture and population size increased. As populations continued to grow, they relied even more on agriculture. (You can support more people with agriculture.) Another change is a shift to becoming more sedentary; you cannot care for crops if you are nomadic.

All of these changes and others led to changes in the pattern of human health and disease. We have been able to find out much about the health consequences of the transition to agriculture by examining the skeletal remains of people that lived before, during, and after this transition. One of the main effects of agriculture is a shift in diet—you can feed many more people with agriculture than by hunting and gathering. (It has been estimated that the world could have only supported about 6 million people as hunters and gatherers,[35] compared with the *billions* that are fed today with agriculture.) However, agriculture does not necessarily mean that people are being fed *better*. When human populations adopted agriculture, their dietary diversity decreased because they focused on a small number of available crops such as wheat, corn, or rice. These are all good foods, but not if your diet does not include sufficient diversity to make up for loss of different amino acids or nutrients. This problem is exacerbated by a reduction of animal proteins in the diet of early agriculturalists. The skeletal evidence shows that the reduction in food diversity resulted in a reduction in health, as determined through analyses of bones and teeth.[36] The transition to agriculture is associated with an increase in dietary deficiencies. For example, iron deficiency anemia is common in early agricultural populations and is seen by the formation of characteristic lesions on the inner upper surface of the eye orbit.

Dental health also declined during the transition to agriculture. We see a marked increase in cavities and tooth loss in a number of early agricultural skeletal samples.[37] Across the world, we see an increase in the percentage of teeth with cavities from about 2 percent in hunting and gathering samples to 5 percent in transitional samples to 9 percent in early agricultural samples. Much of this increase appears to be due to new ways of preparing soft carbohydrate foods, such as gruel, which sticks to teeth and promotes dental decay.

Another health consequence associated with the development of agriculture was a rise in epidemic infectious diseases. Diseases are classified as infectious, which are caused by bacteria or viruses, such as measles, smallpox, and the common cold, and noninfectious, such as heart disease, cancer, and diabetes. Infectious diseases can be spread directly from person to person, such as catching a cold, or through another vector, such as the spread of malaria via mosquitoes. The spread of a disease depends on the number of susceptible individuals in the population; if there are not many individuals, a disease might not spread rapidly. We often characterize disease as endemic or epidemic. Infectious disease in hunting and gathering populations tends to be endemic, where the disease typically is maintained at constant and generally low levels. In an epidemic, the number of cases increases exponentially before declining. An epidemic will happen when more individuals are infected than recover (or die) in a given amount of time.

Population size affects whether an infectious disease will become epidemic. For a number of short-lived microorganisms, hunting and gathering populations are too small to keep the chain of infection going, and we do not see the kind of horrendous epidemics found in later human history, such as the Black Death (bubonic plague) epidemic of the fourteenth century. Because agricultural populations are larger, there are a larger number of susceptible individuals, which keeps an epidemic going for a while. In addition, people in agricultural populations are packed more closely together (as you cannot spread out too far and maintain a farming community), and this increased population density increases the spread of infectious disease because everyone will be in contact with more people.

Other factors also increased the spread of infectious disease in early agricultural populations. One problem with large, sedentary populations is that there is more human and animal waste, which spreads disease. (Keep in mind that this all happened before sewer systems were invented.) In addition, as agricultural societies grew, they became part of larger political and economic networks. Whether contact between humans groups was part of trade and cooperation or warfare and conquest, this contact always allowed the further spread of infectious disease. (We continue to deal with this problem today—disease can spread very fast across the world.) Yet another problem with the origin of agriculture is that humans were in closer and more frequent contact with animals, used for either food or labor, which increased the evolution of diseases to infect human populations. Smallpox, for example, evolved from cowpox.

When we look at the skeletal remains of early agricultural populations, we see ample evidence of the effect of infectious disease and decreased nutrition. For example, growth is affected by both poor nutrition and infectious disease stress, resulting in delayed growth and shorter adult height in early agricultural populations. We also see an increase in dental hypoplasia, which are grooves and pits that form in dental enamel and reflect nutritional and disease stress.

Although there are some advantages of an agricultural life, there are clearly risks as well. The overall health of human populations can be summarized by looking at the life expectancy at birth, the average number of years a newborn child will live. This statistic is derived by knowing the age of death of individuals in the population (a method known as life-table analysis, used by demographers and insurance companies). Age of death can also be estimated from skeletal data when we have a good estimate of the age at death and, given sufficient samples, used to estimate life expectancy at birth in prehistoric populations. Studies of hunting and gathering societies show a range for 20 to 40 years for life expectancy at birth.[38] This does not mean that everyone died before 40 years or age, but rather that there is a large enough number of deaths to infants and young children to keep the average life expectancy low. Although there were older people in these populations, there were not as many as we see in economically developed nations today.

The life expectancy at birth for hunters and gatherers is low compared to recent (2015) values of 71 years for the world and 79 for the United States (although some poorer countries have lower life expectancies in the mid- to upper 40s).[39] For the species as a whole, we tend to live longer than our hunting and gathering ancestors did. Some might assume that the origin of agriculture had something to do with this increase in life expectancy, but this is not the case. In some cases, the life expectancy of early agricultural populations actually decreased. In particular, the survival rate of young children was poor in agricultural populations until relatively recently in history.[40]

It is clear that the origin of agriculture was very much a mixed blessing for humanity, particularly when we consider the initial impact on health and disease. However, over time, there has been improvement in human health and life expectancy; it just did not take place early on when agriculture was developed. In fact, the increase in life expectancy that we take for granted today mostly occurred since the late nineteenth century, the subject of Myth 41.

Myth #40 Civilization has been influenced by extraterrestrials

Status: A popular concept in science fiction is the idea that human evolution was influenced by extraterrestrials. Though fictional, this idea has been popularized as an explanation for the development of early human civilizations, including technological achievements such as the Egyptian pyramids. There is absolutely no evidence to support this view. Human civilizations were developed entirely by humans.

One of my favorite science fiction television shows is *Stargate*, which was developed from the 1994 feature film of the same name. The basic premise of *Stargate* is that in the distant past, a race known as "the Ancients" distributed a series of wormholes throughout space, allowing people and objects to be transported. Thousands of years ago, an alien parasitic race came to Earth and possessed the bodies of humans, taking ancient Egyptians as slaves and spreading them out to other planets to serve their rulers. A rebellion then forces the aliens off Earth. The television series deals with a team of present-day humans who discover the stargates and are involved in both an ongoing fight with the aliens and documenting their influence on the myths and technology of the transported Earth humans. Key to the series concept is the idea that the known achievements of the ancient Egyptians were due to alien intervention. In *Stargate*, the great pyramids of ancient Egypt were built by the aliens and used as landing sites for their spacecraft. I found the series compelling and enjoyed the intersection of modern technology and ancient myths as present-day humans tried to find their place in a much larger and complex universe. However, the fact that I like the stories does not suggest in any way that I place *any* credence in the plot. It is fiction; when the show is over, I return to reality.

The idea that aliens have come to Earth and intervened in our biological evolution and/or cultural development is a popular concept in science fiction. The problem is that people have sometimes confused fact and fiction, advocating that such intervention has indeed occurred. We have no evidence of the actual existence of other intelligent species in the universe that would be capable of space travel. However, if we grant the *possibility* of such species (which I do), then it becomes almost irresistible to wonder if our planet has been visited in the past, and if so, the possible influence of such contact on our own evolution. Of course, this is all speculation. If we want to consider these possibilities from a scientific perspective, we need to have evidence that can be used to test the null hypothesis that there has been no such contact. Conspiracy theories aside, there is no evidence that can be used to reject the no-contact hypothesis. We have no archaeological evidence of spaceships, space suits, antimatter drives, stargates, or other alien technologies. Nor do we have any fossil evidence of ancient visitors from other planets.

The lack of evidence has not stopped speculation about extraterrestrial contact, past or present. One of the most well-known advocates of extraterrestrial influence on human evolution is Swiss author Erich von Däniken, who wrote the bestselling book *Chariots of the Gods* in the late 1960s, as well as numerous later books on the same subject. Von Däniken starts by considering the possibility of extraterrestrial visitors and then using this possibility to explain assorted mysteries of the ancient world, such as the large stone statues found on Easter Island in Polynesia and the large pyramids of ancient Egypt. He also interprets a variety of different myths from the perspective that the gods of such myths were indeed extraterrestrials.[41]

The logic here is that anything that looks mysterious or unexplainable to our eyes is then considered evidence of extraterrestrial contact and influence. How else could primitive peoples have constructed such wonders of the ancient world such as the Great Pyramid and the Sphinx of Giza? Once we have elevated the possibility of alien species to a fact (without any evidence) and assume that our ancestors could not have made such structures, we then conclude that "ancient astronauts" showed our ancestors how to build them (without leaving any evidence of their visit). Archaeologist Kenneth Feder calls this assumption the "Our ancestors, the dummies" hypothesis.[42] This is the idea that ancient humans were too stupid to have built anything as complex as the Great Pyramid or to have carved and moved the large stone statues on Easter Island. Someone had to help them. I have always viewed this assumption as a larger form of a generation gap. We tend to assume everyone in the past is less "advanced" because they lacked our modern technologies and knowledge. For example, in a world where computers are ubiquitous, it is sometimes hard for someone raised on such technology to be able to fathom how airplanes or intercontinental rail systems could have been possible in the past, even though we know they were. Extend this view back over millennia, and we wind up concluding that our ancestors could not possibly built a simple hut, let alone a pyramid!

Thus, the argument for extraterrestrial contact rests in large part on the assumption that our ancestors were incapable of producing the artifacts with which they are associated. However, the scientific evidence shows that this assumption is not valid. As an example, consider the claim that the Polynesians on Easter Island lacked the technology and skills to be able to carve large humanoid statues and then transport them around the island. The hypothesis to test here is that humans could not have built or transported such structures given restrictions on past technology (e.g., no forklifts). The test here would be to see if we can do so. An additional test would involve finding evidence of alien technology. Numerous excavations have found the quarries and stone tools used by the Polynesians to make these statues, but no evidence of more advanced space-age technology.[43] In addition, living Polynesians have been observed carving such statues using old stone tools, and time studies show that while these statues take a long time to make, humans could make them without alien help. The problem of transporting the statues has also been investigated, and several experiments have shown several ways that humans can move them over miles on the island. If humans can accomplish these tasks by themselves, the existence and placement of the statues can therefore not be used as evidence for extraterrestrial assistance.[44]

Another example of scientific evidence against extraterrestrial influence is the Great Pyramid at the Giza plateau in Egypt, near Cairo. The ancient Egyptian pyramids were built as tombs for their rulers who were viewed as incarnations of the gods. Pyramids were important manifestations of the centrality of the afterlife in ancient Egyptian civilization and the importance of having their rulers make a successful journey to the afterlife. The Great Pyramid at Giza dates back over 4,500 years ago and was built for the pharaoh Khufu (known as Cheops by the Greeks), and is the largest of the ancient pyramids. This pyramid has a base length of 230 meters (756 feet), was 147 meters (481 feet) high, and contains over 2.3 million blocks of stone, each weighing about 2.5 tons. In addition to Khufu's pyramid, the pyramid complex also includes two temples, smaller pyramids for three queens, and tombs for many officials, among other monuments.[45] This is truly an impressive feat, and Von Däniken questions the ability of the ancient Egyptians to be able to accomplish the feat given their available technology. How did they carve and move such massive blocks? How did they level the ground? How did they feed the many people needed to construct the pyramid over many years? These are important questions, but Von Däniken views that they are impossible to answer without invoking extraterrestrial help. Von Däniken also suggests that pyramid building appears suddenly without precedent, making the idea of human invention even more incomprehensible and extraterrestrial intervention even more likely.

In reality, we know that pyramid building has a long history in ancient Egypt. Pyramid building did not arise out of nothing, but instead evolved over many centuries. At first, Egyptian pharaohs were buried in simpler temple mounds. Over time, tombs became more complex, and burials were interred in mastabas, which were rectangular one-story structures made of mud. These also became more complex over time, evolving into stepped mastaba, where structures were placed on top of each other, eventually leading to a pyramid shape. Later stepped pyramids were made of stone, and became even more complex over time. The archaeological evidence shows problems arising in construction at times, and solutions taken to perfect various methods, a typical pattern of human invention. By Khufu's time, the art of pyramid construction had developed more fully, reaching a peak with the Great Pyramid, a lasting monument to human creativity and ingenuity, as well as dedication to a single purpose.[46] Contrary to Hollywood productions, the pyramids were not built primarily with slave labor, but by skilled artisans dedicated to ensuring the journey of their rulers to the afterlife.

Long-term excavations of ancient Egyptian pyramids have given us many clues as to how these monuments were constructed, as well as details on the logistics of pyramid building, such as housing and feeding the workers. Additional insight has come from experimental archaeology, where we try to solve various pyramid-building problems by actually performing tasks using the kind of tools that would have been available in the past. Such problems include figuring out how to move the large stone blocks along the sand using ropes, and how to build ramps that wrap around the rising pyramid to move the blocks higher and higher. In each case, there is ample evidence that the pyramids were built by humans, using methods developed over generations,[47] and without the need for alien intervention (or any evidence of such intervention). Our ancestors were not dummies.

How did human civilizations develop if not guided by the hand of aliens? It is important to note that civilizations did not arise instantaneously from nowhere, but developed over time from ever-growing agricultural populations that in turn arose from small agricultural villages. As agricultural populations grew, a number of changes occurred in societies, including labor specialization, social stratification, surpluses of food and labor, and an increased level of formal political control. Eventually, human societies developed into larger state-level groups and cities began to appear. New technologies appeared, including the use of bronze and later iron for making tools. At this point, we often use the term "civilization" to refer to these larger states, which began to appear almost 6,000 years ago. Some of the earliest civilizations appeared in Iraq at the sites of Uruk and Ur, in Egypt at the site of Hierakonpolis, and in South Asia at the sites of Moheno-Daro and Harappa in Pakistan.[48] As with agriculture, civilization developed independently in the Old World and New World.

Of course, the above is a highly simplified bare-bones summary of the development of human civilizations and you should consult world prehistory textbooks to get more detail on the specifics and descriptions of regional variations. However, it does suffice to show that the origin of civilization is an extension of the cultural adaptations that have been apparent since the initial origin of stone tools. Starting with the origin of the genus *Homo* (and perhaps somewhat earlier), our ancestors have relied increasingly on using their brains to solve problems, and their hands and eyes to fashion tools and modify the environment. Some combination of climatic opportunity and population pressure accelerated the development of agriculture and later the rise of civilizations across the world. Whatever the specific details of these transitions, this rich history is a testimony to the abilities of our ancestors and not to the mysterious meddling of aliens from outer space.

The recent increase in life expectancy was due initially to antibiotics

Status: Life expectancy at birth has increased dramatically in some parts of the world and we are likely to live longer than people did even a short time ago in history. This rapid increase in life expectancy has been in large part due to a reduction in infectious disease. It is often assumed that medical advances, such as antibiotics, are responsible for these changes. Actually, infectious disease rates had declined rapidly decades before the first antibiotics were developed. The initial reduction in infectious disease was instead due to public health measures, such as clean water and adequate sanitation.

As was shown in Myth 39, life expectancy at birth has been relatively low for much of our species' existence. Hunting-gathering populations typically had life expectancies in the range of 20 to 40 years at birth, a number that did not increase with the initial development of agriculture. Even in more recent historical times, the life expectancy at birth had not risen all that much. In the United States for example, the estimated life expectancy at birth from the late eighteenth century through much of the nineteenth century was in the upper 30s to low 40s for white males and females. By the end of the nineteenth century, life expectancy at birth increased to the upper 40s for white males and females.[49]

Starting in 1900, data on life expectancy at birth have been estimated every year for the entire US population. These values are shown in Figure 4.1 for males and females for every year between 1900 and 2011. The overall trend since 1900 has been an increase in life expectancy at birth. In 1900, the average newborn male had a life expectancy at birth of 46 years and the average newborn female had a life expectancy at birth of 48 years.[50] By 2011, the life expectancy at birth had risen to 76 years for males and 81 years for females.[51] This increase has not been unique to the United States; the more economically developed nations of the world have typically seen impressive gains in life expectancy throughout the

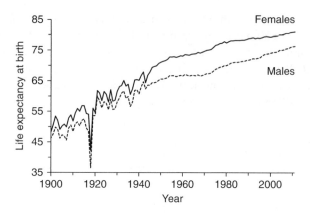

Figure 4.1 Changes in life expectancy at birth in the United States from 1900 through 2011. The solid line is for females and the dashed line is for males. Data for 1900 through 2009 from Arias (2014) and data for 2011 from Hoyert and Xu (2012). Life expectancy at birth has risen over time, except for a major decline in 1918 owing to deaths from an influenza pandemic.

past century. A number of countries have even higher life expectancies at birth, in excess of 80 years.[52]

Figure 4.1 also shows two other facts. First, for any given year, the life expectancy at birth is higher for females than males, something found in most human populations. Second, both male and female curves show a noticeable decline in life expectancies for the year 1918. This decline is a consequence of the 1918 influenza pandemic. (A pandemic is a widespread epidemic.) Because so many young people died that year due to the flu, the life expectancy at birth, which is based on the age of death, plummeted. Excluding 1918 and some minor fluctuations from year to year, the general trend of increasing life expectancy is clear. This means that the average child born in 2011 was expected to live 32 years longer than the average child born in 1900 did—a 68 percent increase.[53]

Why are people today living longer on average than a century ago? The increase in life expectancy is due to a reduction in annual mortality. The decline in annual death rate in the United States is shown in Figure 4.2 from 1900 to 2011, showing a very clear and regular decline during the twentieth century. When we look more closely at different causes of death, we see that much of this decline is due to a reduction in deaths due to infectious disease. In 1900, the death rate due to infectious disease in the United States was 797 per 100,000 people, declining to 59 deaths per 100,000 in 1996—a 93 percent reduction overall (although there was an increase in 1918 due to the flu and then between 1981 to 1995 because

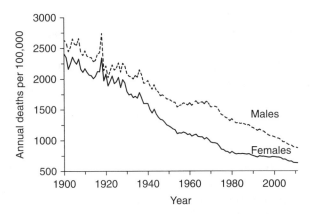

Figure 4.2 Annual death rates (deaths per 100,000) in the United States from 1900 through 2011. Death rates have been age-adjusted based on the 2000 Census to control for variation in age composition over time. Data from the Centers for Disease Control and Protection web page at https://data.cdc.gov/NCHS/NCHS-Age-adjusted-death-rates-and-life-expectancy-/w9j2-ggv5 (accessed June 11, 2016).

of HIV). The reduction in deaths due to infectious disease during the twentieth century was particularly marked in the very young (less than 5 years of age).[54] The decline in infectious disease deaths among infants and young children explains the increase in average length of life. Because life expectancy at birth is a statistic based on the age distribution of those who died in a given year, a large number of deaths of infants and young children will lead to a lowering of the life expectancy of a newborn child. When fewer infants and young children die, the average newborn has a higher probability of living longer, and life expectancy at birth increases.

The reduction in infectious disease deaths in the United States during the twentieth century is also seen by looking at changes in the leading causes of death. During the nineteenth century, the death rates of infectious disease were high, particularly during the shift in population from rural to urban areas where overcrowding, poor sanitation, inadequate waste disposal, and contaminated water led to epidemics of a number of infectious diseases. Although improvements began in the late nineteenth century, infectious disease was still the top killer in the United States at the start of the twentieth century. In 1900, the top three leading causes of death were influenza/pneumonia, tuberculosis, and diarrheal diseases/enteritis. (Influenza and pneumonia are typically lumped together in mortality reports because pneumonia is often caused by flu.) These three leading causes of death were all infectious diseases. Chronic noninfectious diseases ranked lower, including heart disease (fourth leading cause) and cancer (eighth leading cause).[55] These trends are in part a reflection of risk by age group; heart disease and cancer disproportionately affects older individuals and in 1900 many more people died earlier in life due to infectious disease. Many did not live long enough to die eventually from heart disease or cancer.

Since that time, the leading causes of death in the United States have changed. The top three causes of death in 2010 were heart disease, cancer, and chronic lower respiratory diseases. Of the top 10 causes of death in 2010, only one, influenza/pneumonia, is an infectious disease (at number 9). Although still prevalent, the death toll from infectious disease has clearly decreased and the death rate due to chronic noninfectious causes has increased proportionately; the two leading causes of death, heart disease and cancer, accounted for almost half (48 percent) of all deaths in the United States in 2010.[56]

All of these changes are often summarized as part of a shift in health known as the epidemiologic transition, a term used to describe changing patterns of disease mortality.[57] As a population undergoes modernization and develops a public health infrastructure, death rates due to infectious disease drop, particularly in the younger age groups. People live longer, which leads to an increase in life expectancy. As people age, their eventual demise is more likely to be from a chronic noninfectious cause of death, such as heart disease or cancer, than from an infectious disease (in relative terms).

The decline in deaths due to infectious disease, particularly those that affect the young, is at the heart of the epidemiologic transition. What happened? When we think of battling infectious diseases, which are caused by viruses and bacteria, we think of modern medicine, which includes vaccines and antibiotics. Vaccines are used to protect against the development of viral diseases and antibiotics are used to counter bacterial diseases. It is therefore tempting to suggest that the early twentieth-century reduction in infectious disease was due to the development and use of vaccines and antibiotics. Although these drugs have been quite useful in continuing our battle against infectious disease, they were not responsible for the *initial* reduction of infectious disease during the epidemiologic transition.

Vaccines are now available for a number of infectious diseases, but the early use of vaccines was more limited. Edward Jenner developed the smallpox vaccine in the late 1790s and its international use led to the eradication of smallpox worldwide by 1977. However, the bulk of vaccines that we have today, including vaccines for flu, measles, mumps, and other diseases were developed after the middle of the twentieth century, after the reduction in infectious disease deaths had already started. What about antibiotics? The first antibiotic to be used on humans was penicillin, discovered in 1928 and first used to treat an infection in 1942.[58] This occurred long after the start of the reduction in deaths due to infectious disease, which was well under way by the beginning of the twentieth century (refer back to Figure 4.2). Although vaccines and antibiotics have helped us continue to lower infectious disease, they were not responsible for the first reductions in infectious disease.

What then led to the beginning of the epidemiologic transition? It was due to medical science, not in the form of a specific type of medicine, but instead a change in policies and engineering that resulted from the application of a newly discovered principle of medicine—the germ theory of infectious disease. Developed by Louis Pasteur and others, germ theory showed that many diseases were caused by microorganisms that could spread disease. Some of these microorganisms spread through the air, some through water, and some by direct transmission from one person to the next. Knowledge of how disease spreads led to a number of changes that led to a decline in infectious disease at the end of the nineteenth century and continuing into the twentieth century. Changes were introduced following the establishment of public health departments at local, state, and federal levels. These agencies promoted actions to reduce the spread of infectious diseases through improvement of infrastructure.

Two of the major changes in infrastructure were providing for clean water and the development of sewer systems, both of which cut back on the transmission of infectious microorganisms. In addition, new methods of waste disposal were introduced, laws and regulations were passed regarding food safety, and educational efforts were geared toward the importance of proper hygiene.[59] As a consequence of these (and other) changes, infectious diseases declined quickly, leading to the changes in life expectancy and death rates shown in Figure 4.1 and Figure 4.2. Note again that the major reductions in infectious disease occurred *before* the widespread use of vaccinations for a variety of diseases or the initial use of antibiotics. These marvelous medical advances have helped to *continue* to reduce the threat of infectious disease. Smallpox has been eradicated and some predict that polio will soon be eliminated as well. The incidence of other diseases, such as measles, has been greatly reduced relative to a century ago as the result of continued vaccination. Although medical experts had once been a bit too optimistic in predicting the conquest of all infectious disease, the overall pattern of health has improved in many countries.

Still, it would be naïve to think that infectious disease has been conquered. For one thing, although many disease rates have gone down, some are still relatively high. Although influenza/pneumonia is no longer the number one killer in the United States, it is still in the top 10 leading causes of death and had killed over 50,000 people in the United States in 2010 (accounting for 2 percent of all deaths).[60] It is also important to remember that the reduction in infectious disease has to date been greater in the more economically developed regions of the world. Elsewhere, infectious disease is still unfortunately high, such as in sub-Saharan Africa where the top four causes of death in 2012 were due to infectious diseases (HIV/AIDS, lower respiratory infections, diarrheal diseases, and malaria).[61]

There is a danger in assuming that infectious diseases will naturally continue to decline even in the more economically developed nations because the microorganisms that cause infectious disease continue to evolve. Bacteria and viruses mutate, and changing environmental conditions can select for new strains of diseases. A number of infectious diseases that are found in other animals can continue to adapt to human hosts. Given how human populations are all interconnected globally, new diseases often have the potential to spread across wide areas, particularly in this age of rapid air transport around the world.[62] One of the most notable examples of an emergent infectious disease is AIDS, caused by the HIV virus. Other emergent diseases in recent times include Legionnaire's Disease and Ebola. In addition, a number of infectious diseases are considered "reemergent," where a disease that had previously been reduced starts to increase in incidence. This reversal often happens because of bacteria evolving resistance to antibiotics. One example is tuberculosis, a respiratory disease that, as noted earlier, was the second leading cause of death in the United States in 1900. By the middle of the twentieth century, antibiotics had been developed to battle tuberculosis. Although successful, some strains evolved resistance to antibiotics and the incidence rates went up in the last two decades of the twentieth century.

Infectious disease has not been conquered and rates can go up when conditions change, as illustrated by the fact that infectious disease deaths in the United States actually increased somewhat in the 1980s and early 1990s. This increase was due in part to the emergence of AIDS and in part to an increase in deaths from influenza/pneumonia among the elderly.[63] Such examples show us that we must be ever vigilant in our fight against infectious disease. It is not a threat that will be dealt with in a single war, but rather a series of continuing battles. Continued success against infectious disease requires maintenance of adequate infrastructure (such as clean water and sewer systems), control of disease-carrying organism (e.g., mosquitoes), and widespread vaccination, among other efforts.

There are three distinct shades of human skin color

Status: Although we often use discrete categories to describe human skin color ("black," "brown," "white"), skin color actually shows a continuous distribution in the human species, ranging from very dark to very light with every shade of pigmentation in between. The geographic distribution of human skin color shows the results of past evolution that has favored darker and lighter pigmentation in different environments.

We tend to categorize and group objects into a small number of categories when possible, part of our propensity to make sense out of the natural and social world. Some objects are easily placed into different, nonoverlapping categories, such as money, which can be organized into coins versus paper money. However, many of the things we want to classify are not always easy to place into different categories. As an example, consider human height. In any large sample of human adults, we will see a range of variation from the very shortest individuals to the very tallest. When we describe height in a casual sense, we use groupings such as "short," "medium," and "tall" for convenience, but we also realize these groupings are arbitrary. For example, where is the cutoff between "short" and "medium?" In addition, why should we limit ourselves to only three groups? Why not add other groups such as "medium-short" or "very tall?" We use these rough groupings as a crude approximation to the reality of height. We do so realizing that these categories are crude and do not actually think that a population is only made up of people with three different heights. (If it were, making clothing would be much simpler.) There are many other examples of how we try to reduce continuous variation into a small number of distinct groups. Take socioeconomic status for example; we might use terms such as "poor," "middle class," and "rich" as convenient groupings for some purposes, but we also know that there are not really only three different levels of income.

Skin color is one of those physical traits that are continuous, but we still often try to classify individuals into one of a small number of groups, typically using terms such as "black," "brown," and "white." The continual use of a three-group system of classification makes it easy for us to fall into the trap of thinking that these terms are accurate descriptions of the variation in human skin color and that people only come in three distinct shades of pigmentation. Although we can see variation within groups, such as some light-skinned people having lighter skin than others, or some dark-skinned people having darker skin than others, the three-shade system tends to work well in many situations. In many parts of the United States, you can stand on a street corner and classify people as they walk by, easily placing them into categories of "white," "brown," and "black." It then seems natural to extrapolate this experience to consider *all* of humanity as being made up of people of three distinct shades of skin color. Or, if you have a keen eye toward variation, perhaps you might add several groups.

Of course, part of the reason that it is often easy to classify people as black or white in the United States lies in the nation's specific history. Large numbers of people have ancestors who came from northwest Europe, an area populated by very light-skinned people, and a large number have ancestors who were enslaved and taken from the western part of Central Africa, an area populated by very dark-skinned people. If people walking down the street were a proportional representation of people from around the world, we would see a different story. Imagine a large group of people from all around the world lining up across a football field, arranged by their skin color. You would see very dark people at one end of the field and very light people at the other end of the field, with a continuous range of pigmentation across the field. You could easily see that some people are darker than others are, and could label the ends of the field as "black" and "white," and those at the 50-yard line as "brown." However, you would not find it easy to identify the exact number of shades that you see, or give a precise position on the playing field for the cutoff from one shade to the next. At best, your attempt at classification would be arbitrary.

How do we know this? Has a random sample of humanity ever been lined up on a football field? Not that I know of. However, we do have statistics on skin color from around the world and we can use these statistics to perform the football field experiment using a simple graph. Anthropologists and human biologists use a very specific way of measuring human skin color. Although skin color was once measured by comparing someone's skin to a set of tiles (much as you might match paint samples), since the 1950s we have relied upon devices known as reflectance spectrophotometers. Here, light at different wavelengths is shined on someone's skin to measure how much light is reflected back. Someone with very light skin will reflect more light than someone who has darker skin.

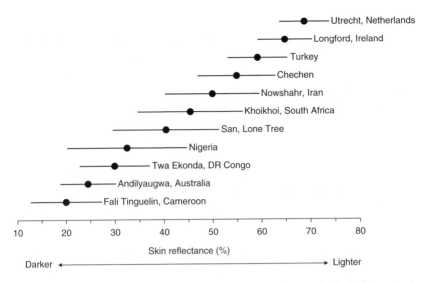

Figure 4.3 Variation in skin color for 11 human populations (males). Skin color is measured as the percentage of red light (wavelength of 685 nanometers) reflected off the skin; lighter skin reflects more light. The dots for each population indicate the average skin reflectance in the sample. The lines indicate two standard deviations below and above the mean, a range that includes slightly more than 95 percent of the population, and gives a good visual index of variation within each population. Although some human populations are very dark, some are very light, and some are in between, the distributions of skin color overlap other populations; there are no distinct breaks or shades. Data sources listed in Relethford (1997, 2000). Populations were picked from a larger list, sampling one population approximately every five percentage points of reflectance.

To get around problems of measurement being complicated by tanning, we measure skin reflectance at the upper inner part of the arm, which is relatively unexposed.

There have been over 100 or so studies of human skin reflectance from around the world. Figure 4.3 shows data from males in 11 different indigenous populations from Africa, Eurasia, and Australia. Each of these populations was measured using the same type of machine, which measured the amount of red light reflected off the skin. (Red light works best here because it reflects the variation in melanin, the pigment that affects levels of pigmentation.) Each population shows the mean (average) skin reflectance as a dot on the graph and the range from plus and minus 2 standard deviations from the mean. A standard deviation is a measure of variation (see any statistics textbook) that we need not define in detail; the important thing is that the range defined by plus and minus 2 standard deviations accounts for more than 95 percent of the sample, giving us a graph of the level of variation in the population. For example, we see that the Fali Tinguelin, a group living in northern Cameroon, is very dark as the average skin reflectance is 20 percent. We also see that there is variation *within* this group—some of the Fali Tinguelin are darker than average and some are lighter than average.

We now turn to the next population on the graph, the Andilyaugwa aborigines in Australia. They are slightly lighter than the Fali Tinguelin *on average*, but it is clear from the graph that the skin color distributions for these two populations overlap, so that some Fali Tinguelin are actually lighter than some Andilyaugwa. Looking at all 11 populations on the graph shows that each group overlaps in skin color with its two adjacent groups. If you compress all of this variation, you would get a single straight line ranging from the darkest Fali Tinguelin to the lightest resident in the Netherlands, with no breaks. Although we could call one end of this line "black" and the other end "white," there are no naturally occurring cutoff points to be able to count a specific number of shades or to determine their cutoff points. Human skin color does not come in distinct and nonoverlapping shades.[64]

The continuous nature of human skin color raises some interesting questions regarding the evolution of human skin color. Most physical traits in humans do not show such a wide range of differences between people in different parts of the world, which is a reason people have used skin color for racial classification. The difference between humans across the world is relatively much greater for skin color than most physical or genetic traits, which typically show much less differentiation between groups.[65] This range of variation typically suggests a trait that has been shaped by natural selection operating in different ways across environments. What then is the pattern of skin color variation when looked at in terms of environmental correlation?

Quite simply, skin color tends to be darker in indigenous human populations that live at or near the equator and humans tend to be increasingly lighter with increasing distance away from the equator.[66] This pattern is seen very clearly in Figure 4.4, which shows the average skin color as a function of distance from the equator in 107 human populations in the Old World. Although the correspondence is not perfect, it is a very strong relationship and shows that skin color is lighter away from the equator, both north and south.

The strong relationship between human skin color and distance from the equator gives us clues about the evolution of skin color in humans. It turns out that ultraviolet radiation also varies by latitude, with the

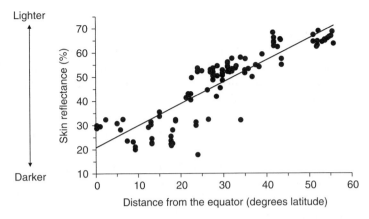

Figure 4.4 The geographic distribution of human skin color in 107 Old World populations (males). Skin color is measured as the percentage of red light (wavelength of 685 nanometers) reflected off the skin; lighter skin reflects more light. The dots indicate the distance from the equator and average skin color for each population. The solid line is the linear regression line showing skin reflectance as a function of distance from the equator, and is the best fitting line that can be drawn through the points. Data sources listed in Relethford (1997, 2000).

strongest levels occurring at and near the equator and decreasing with farther distance away from the equator. The fact that both skin color and ultraviolet radiation are correlated with distance from the equator, and the fact that ultraviolet radiation can have a number of harmful effects, suggests that dark skin evolved in our early African ancestors as protection against excessive ultraviolet radiation. Heavily pigmented skin is a form of natural sunscreen. Current thinking is that when early members of the genus *Homo* began spending more time on the open savanna, there was an increase in sweat gland density to adapt to the heat and an associated loss of hair. Early humans would have had somewhat pale skin, as do African apes have under their fur. As such, our early ancestors would have been subjected to the dangers of too much ultraviolet radiation, which would be a problem near the equator once our ancestors came out of the shaded forest into the open savanna. According to this model, there would have been selection for dark skin to protect against the harmful effects of ultraviolet radiation.

What are these harmful effects? A number of hypotheses have been suggested and debated.[67] Sunburn is one outcome. Although it can be uncomfortable, in today's world we often tend to treat sunburn as a minor inconvenience. However, it could have been more serious for our ancestors as severe sunburn can lead to infections. Excessive exposure to ultraviolet radiation also increases the risk of skin cancers, although the potential evolutionary impact has been rejected by some because such cancers might have their greatest impact later in life, after an individual has already passed their reproductive years, and therefore not subject to selection. However, it has also been suggested that there may have been enough of an effect during the reproductive years of our ancestors to have an effect on selection.[68] Ultraviolet radiation has been suggested to affect a person's folate levels and these reduced folate levels can cause neurological problems in developing fetuses as well as hamper sperm production in men.[69] Another possible factor in the evolution of dark skin color is the function of the skin as a permeability barrier that prevents drying out and that resists infection. Darker skin performs these barrier functions better and is more resistant to infection. If so, then the evolution of darker skin color in the tropics might not be related to ultraviolet radiation levels, but to heat and aridity, which also vary by latitude.[70]

These potential effects are not mutually exclusive. It is possible that the initial development of darker skin color in our ancestors was due to several factors and not any single cause. What is clear, however, is that as humans dispersed out of Africa moving further away from the equator, something changed and we see a gradient in skin color in the world today. Why did skin color change? As we move further from the equator, the dangers of living at or near the equator diminished and consequently the protective value of darker skin diminished. At the same time, the gradient in skin color suggests that there must have been an increase in the adaptive value of lighter skin in populations far from the equator. Several ideas have been proposed to explain the evolution of lighter skin at higher altitudes. One possibility has to do with vitamin D, a needed nutrient for proper bone development and other biological functions. Although some foods have high levels of vitamin D, most humans throughout history and prehistory have received vitamin D through biochemical synthesis stimulated by ultraviolet radiation. According to this hypothesis, as humans moved further from the equator, the dangers from having too much ultraviolet radiation decreased and the problem shifted to not having enough, a problem that became greater with increasing distance from the equator. Not everyone agrees with this hypothesis, suggesting that sufficient levels of vitamin D can be synthesized at northern latitudes even in dark-skinned people.

Regardless of the specific factors involved at different latitudes, the geographic distribution of skin color shows a balance between different selective pressures over space, resulting in a continuous range of variation, with no distinct shades.

Biological race is useful for understanding human variation

Status: Human variation is often discussed in terms of the biological concept of race, the idea that our species can be broken down into a set number of discrete races that can be identified using genetic and/or physical traits. At times, race has been considered as a unit of evolutionary change (subspecies) and at other times as a descriptive term referring to an aggregate of populations in a geographic region (geographic race). Neither of these approaches has been useful in describing and analyzing human biological variation. Genetic variation exists among human populations and is geographically structured (populations are typically most similar genetically to their neighbors), but the key features of global genetic variation in humans is not captured by the biological race concept. Race is a label and not a biological reality.

When people discuss human variation, it is often from the viewpoint of race, an approach than stands in contrast to the research conducted today in human biological variation. For example, much of my current research look at global variation in cranial measures, and I have also conducted research on global variation in skin color, red blood cell genetic markers, and DNA markers. The common element of these studies is the analysis of patterns of variation from an evolutionary perspective, focusing on the relative impact of mutation, selection, drift, and gene flow on genetic differences between local human populations. None of my work uses the biological concept of race, except for historical narrative and critiques of the concept. Given the assumed importance of race as a tool for understanding human variation, why not use it? The simple answer is that the biological race concept is not useful in describing and analyzing patterns of human biological variation.

Discussions of race are complicated by the range of definitions and applications of the term. What do we mean by race and how useful is it as a concept for describing and analyzing human diversity? The term "race" has been used for many different things, both biological and cultural. Race is a form of classification, but the criteria used for labeling and identifying races vary quite a bit. Race is often described in terms of ancestry, geography, nationality, cultural identity, religion, and language, among others. Not all of these definitions are consistent and may not overlap with other definitions. Because we have a variety of cultural and biological definitions of race, it is possible that two people can talk to each other about race and actually be talking about quite different things.

This myth deals specifically with biological concepts of race. Here too, there are different definitions that have been used over time, contributing further to confusion.[71] The traditional definitions of race in biology focus on divisions of a species that are distinct from each other biologically. Often, race has been considered equivalent to a subspecies, a grouping that represents early stages in the formation of a new species. Here, race is taken as a unit of evolutionary change, implying that races are evolving more or less independently. An example of this approach is in the work of anthropologist Carleton Coon, who proposed in the 1960s that there were five races of early humans that evolved into five races of modern humans, a view that has long been discredited.[72] Although the concept of subspecies works well for a number of species, it does not do well when applied to humans.[73] The extent of biological variation in the human species and its distribution argue against dividing humanity into different subspecies, and biologists and anthropologists place all of humanity in a single subspecies—*Homo sapiens sapiens*.

All biological concepts of race propose that the human species can be classified into a set number of discrete groups based on biological characteristics. The problem with this approach is that there has never been consensus in the scientific community regarding the number of human races. Carl Linnaeus, the sixteenth-century founder of biological taxonomy and classification, identified four "varieties" of humanity—Africans, Europeans, Asians, and Americans (the last referring to Native Americans). Others agreed about Africa and Europe, but lumped Native Americans in with Asians, coming up with three races. There was also debate over where to place South Asians (Indians) or whether to consider them an additional race. Similar arguments took place over the placement of Australian Aborigines, Pacific Islanders, and other groups. Some subdivided the races further; for example, splitting Africa into two races or distinguishing between East Asian and South Asian races. By the mid-twentieth century, there were still arguments about how many human races existed. For example, geneticist William Boyd identified six races (including placing the Basque of Europe into their own race), anthropologist Carleton Coon identified five races (including two in Africa), and anthropologist Stanley Garn identified nine races.[74] Historically, some argued for more races and some for fewer races, a clear sign that there is an element of subjectivity in picking an exact number. Darwin recognized this problem in his 1871 book, *The Descent of Man*, where he noted that the number of races in his time had been estimated to be anywhere from two to over 60. Darwin states that this shows that races "graduate into each other, and that it is hardly possible to discover clear distinctive characters between them."[75] The problem of not being able to agree on a specific number is evidence that the nature of human variation does not allow agreement on where to place the cutoff points biologically or geographically. The situation would be different if humanity was divided into a number of isolated groups around the planet, but that is not our history. Our interconnectedness with other groups means that we cannot draw distinct lines with which everyone can agree.

Another problem with the view that the human species is broken down into a set number of discrete groups is the relatively low level of genetic differences between human groups. Human populations do not show the level of geographic and genetic isolation expected from diverging subspecies. For example, when we compare genetic differences between human populations with genetic differences between African ape populations, we find less genetic difference between human groups, even though we are spread out over an entire planet compared with the smaller geographic range of ape populations.[76]

Another example of how human genetic diversity does not fit racial models comes from the observation of the nested nature of genetic diversity in human populations. Take the common assumption that different geographic regions (such as sub-Saharan Africa, Europe, East Asia, and others) represent different races. If so, then we would expect patterns of genetic diversity to support this division into races. In actuality, the genetic variation we observe in DNA markers shows that genetic diversity outside of sub-Saharan Africa is a subset of the genetic diversity within sub-Saharan Africa, and not separate from it. Using this variation as a guide, we cannot classify sub-Saharan Africans as a separate race, because all other human populations are a subset of sub-Saharan Africans and would belong to the same group. If we tried to force this variation into a racial model, we would wind up considering populations in Europe and Asia as subraces within an African race, a view counter to traditional classifications. The nested nature of human genetic variation does not agree with traditional models of race, which proposes mutually exclusive groups.[77]

Although anthropologists today do not equate race with units of evolutionary change such as subspecies, there has been debate about the usefulness of defining "geographic races," which divide our species into groups based on broad geographic regions, such as sub-Saharan Africa and Europe. This approach recognizes that biological variation *does* exist in our species and it *is* most often structured geographically. People from different parts of the world tend to look different from one another in terms of both physical traits and genetic markers. In general, the farther apart two populations are from each other, the more different they are. No one would be surprised to learn that the frequencies of genetic markers of people in England and in Italy are more similar to each other, on average, than either is to the frequencies in Australian Aborigines. There are exceptions, of course, but the general principle of geography as a prime determinant of patterns of variation holds for blood markers, DNA markers, and cranial measures, among others.[78] We tend to be most similar to neighboring groups (ignoring recent migration) than those farther away. However, the idea of geographic race is not useful for describing this geographic structure.

No one denies human biological variation. The argument is about the best way to *describe* this variation. Sometimes we use broad labels for convenience; for example, a geographically based name, such as "East Asian," gives us an idea of the location of a group. The problem lies when we take our labels and consider them fixed in nature rather than just being labels of convenience. Historically, the race concept has often been used to refer to discrete and nonoverlapping groups that are easily identified, which is not the case with human variation. Assignment of the geographic label "East Asian" does not mean that there is a separate East Asian race.

As an analogy, consider variation in human height. For convenience, we might describe the range of variation in human height in terms of three groups—short, medium, and tall—but this does not mean that three different groups of people exist that fit nice and neatly into these categories. We recognize that height is continuous in nature and any attempt to delineate where "short" ends and "medium" begins will be subjective. Further, we know that when we place people into these categories that we will wind up with some short people who are closer in height to some medium-sized people than they are to other short people. The decision to apply three labels for height is arbitrary (we could just as easily propose five or another number) as are the cutoffs for determining who belongs in each height group.

For centuries, scholars and the public have recognized that people look different from other parts of the world, an observation that we can extend even further using genetic data today. However, the nature of much human variation is continuous, making the decision as to where to draw the line separating one race from the next an arbitrary one. As such, there has never been consensus among scientists on the exact number of human races, because, as in the height example, it has a subjective component resulting from cramming continuous variation into a small number of discrete groups. Racial classification assumes that humanity lives in geographically distinct groups, but we do not. If we start off with races based on populations in Northwest Europe and East Asia (regions that people often think about when talking about Europeans and Asians), what do we do about those groups that live in between these two localities, such as Central Europe, East Europe, and Central Asia? What about other areas in Eurasia, such as South Europe, Southwest Asia, South Asia, and Southeast Asia? What about populations within each of these regions, such as Italy and Greece, and populations within these subregions, such as southern and northern Italy? Depending on how we want to divide humanity, the number of geographic races might range into dozens or more.

Many anthropologists have questioned the utility of geographic races because of the indeterminacy in picking the number of races, as expected when we try to break down continuous variation into a set of discrete, nonoverlapping groups. (Recall the example about different shades of skin color from the previous myth.) Others have argued that despite these flaws there are times when a broad breakdown of humanity is useful, even if the dividing lines are arbitrary. The problem here is that such an approach contrasts with traditional views of race that emphasize discrete groups. There are occasions when it is useful to examine broad patterns of variation by using large continental groupings as the unit of analysis, but these regional groups should not be referred to as races as the history of the biological race concept implies something different than a simple label. This is not a case of politically correct semantics, but refers to underlying differences in assessing variation. The biological race concept is not useful for describing human biological variation.

Myth 44 All African Americans have the same genetic history

Status: Early genetic studies of African Americans attempted to estimate the amount of European ancestry in the African American gene pool, giving a rough estimate of about 20 percent. However, the idea that a single number can summarize the genetic history of all African Americans is incorrect. Later studies showed considerable variation in European ancestry depending on the specific population, and even greater variation between individuals within populations. Some African Americans have virtually no European ancestry and others have quite a bit.

One common problem with looking at human variation is the tendency to focus on differences *among* groups, with less attention to variation *within* groups. Statements such as "They look different than us" and

"They all look alike" are examples of paying more attention to differences among groups and less on variation within groups. This way of looking at the world is embodied in any group stereotypes if we consider any member of a group as representative of the entire group. There is sometimes an unfortunate tendency to see different groups of humanity as being homogeneous biologically.

A focus on types is likely to lead us to miss variation within a group. An example of this problem is looking at European ancestry in African Americans. Starting in the 1600s, hundreds of thousands of Africans, mostly from West Africa and West-Central Africa, were enslaved and brought to the United States. Over time, there was genetic admixture from Europeans into the African American gene pool. Early on, this admixture was primarily due to enslaved African women having babies fathered by European males. There has likely been some additional recent admixture because of changes in laws and attitudes regarding intermarriage.

By the mid-twentieth century, genetic studies of African Americans addressed the question of how much European ancestry was in the African American gene pool. The earliest studies used a relatively simple but informative way to estimate ancestry. Suppose we have a mixed population that has ancestry from populations A and B. Now, suppose that the frequency of a given allele is 80 percent in population A, 40 percent in population B, and 60 percent in the mixed population. Because the mixed population had a value exactly halfway between the values for A and B, we would conclude that its ancestry is half from A and half from B. By looking at the allele frequencies in mixed groups relative to ancestral groups, we can estimate the total amount of ancestry. In actual practice, we need to base such estimates on a large number of different genetic markers to get a more accurate average.

This basic method was first applied in the 1950s and 1960s to genetic data from African Americans as compared with Europeans and West Africans. A number of early studies gave estimates of about 20 percent European ancestry (and therefore 80 percent African ancestry, excepting cases where other groups were involved, such as a small contribution of Native American ancestry in some cases). Keep in mind that these numbers reflect the total amount of ancestry accumulated over many generations, and not the amount of mixing that would take place in any given generation (a different estimate).[79]

Since that time, complex methods of ancestry estimation have been developed and extended to a large amount of DNA markers. Those studies with the largest number of genetic markers give average estimates of about 13 to 19 percent European ancestry in African Americans.[80]

However, this range is a composite based on a wide range of populations and individuals with different ancestral histories. Closer analysis of the genetic history of African Americans shows that no single number can be applied to *all* African Americans.

A good example of the variation in ancestral history is seen in the analyses of DNA markers performed by anthropologist Esteban Parra and colleagues, who estimated European ancestry in 12 African American populations (both cities and rural areas) in the United States.[81] They found that the amount of European ancestry ranged from 4 percent in the Gullah of rural South Carolina to 23 percent in New Orleans. Some of the variation in several South Carolina populations reflects urban and rural differences, with more European ancestry found in the urban sample. In any event, there is clearly variation. There is no single African American genetic history any more than there is a single European American or Asian American genetic history.

Parra's studies also showed that we see different ancestral histories depending on whether we look at maternal or paternal ancestry. They used genetic markers from mitochondrial DNA to estimate maternal ancestry (because mitochondrial DNA is passed on only through the mother) and from Y chromosome DNA to estimate paternal ancestry (because the Y chromosome is passed on from father to son). Again, the proportions of European ancestry varied from population to population, but consistently showed more European ancestry in the paternal line than in the maternal line. For example, in Columbia, South Carolina, the estimated amount of European ancestry was 24 percent from Y chromosome DNA and less than 3 percent from mitochondrial DNA. These results are consistent with the early history of African American populations where male European slaveholders fathered children with enslaved African American women. Although census data show that in recent decades it has been more common for black men to marry white women than the reverse,[82] this trend has so far not had a noticeable effect on sex-specific ancestry estimates.

It is important to remember that all of the estimates of European ancestry discussed so far refer to *average* estimates within populations. For example, Parra and colleagues found that the amount of European ancestry was 18 percent in Columbia, South Carolina. This is an average figure based on the allele frequency in the entire sample and does not mean that every African American in Columbia had 18 percent European ancestry (any more than when I say that the average height of women in the United States is 1.63 meters (5 feet 4 inches) does not mean that all women are 1.63 meters tall).

With the right information, we can estimate ancestry on *individuals* within a population. Parra and colleagues did this for the African American population in Columbia and found some very interesting results.[83] They found that almost half of African Americans in Columbia had between 0 and 10 percent European ancestry, and only a bit more than 10 percent had between 11 and 20 percent European ancestry. Some individuals had more European ancestry, including a small proportion that had *more* than 50 percent European ancestry, and even a few that had more than 90 percent European ancestry.

These results have been confirmed by recent studies of massive numbers of DNA markers. Katarzyna Bryc and colleagues examined over 250,000 genetic markers to estimate European ancestry in 365 African Americans from across the United States. They found a median value of 19 percent European ancestry. However, this sample also showed extensive variation among individuals. Half of the sample had between 12 and 28 percent European ancestry. Further, some African Americans had less than 1 percent European ancestry and some had more than 99 percent European ancestry.[84] Another DNA-based study of African Americans also found an extensive range of West African and European ancestry.[85]

It is very clear that no single number can possibly describe the genetic history of all African Americans. All we really learn from an average estimate for all African Americans is that a proportion of their ancestry typically derives from Europe, primarily from European slaveholders in colonial times. The actual amount of European ancestry varies by population and by individuals, all of which have different ancestral histories. We cannot accurately capture this history with a single number, but have to include measures of variation in any ancestry study. Some individuals have virtually no European ancestry and others have almost entirely European ancestry. This wide range calls into question the notion that African American ancestry is best defined exclusively in biological terms. Instead, whether one is African American is a matter of cultural identity, which does not always match up with our ideas of genetic ancestry (a theme explored in detail in the next myth).

As noted by Bryc and colleagues, among others, there are important implications for understanding that individuals that self-identify as African Americans can have quite diverse genetic backgrounds. Some of these implications concern medical treatment, as rough measures of ancestry are often used in studies of public health and medicine to explore differences in relative risk of disease as well as differences in pharmaceutical treatment. If there is a difference in diagnosis or treatment based on an assumption that one's self-identified racial identity is a reflection of a single pattern of genetic ancestry, then we can make mistakes because not every member of the culturally defined group has the same genetic history. Racial classifications often obscure underlying genetic variation. Genetic risks for disease may soon be identified based on individual DNA studies and not on assumptions based on average patterns in culturally defined groups.

Myth 45 Genetic ancestry is the same thing as cultural identity

Status: Discussions of race often confuse genetic ancestry with cultural identity. Sometimes these concepts overlap and sometimes they do not. We cannot always assume that knowing one's genetic ancestry will necessarily tell us something about their cultural identity, or the reverse.

As noted in previous myths, discussions of race are often confusing out of context because it is a concept that is sometimes defined by biology and sometimes by culture. The problem of understanding what is meant by race increases because the biological and cultural categories sometimes overlap, although not always. To some, race is a term that describes one's genetic ancestry and to others it refers to cultural identity. In some cases, these two concepts agree with each other and in other cases they do not.

An example helps bring these abstract ideas into focus. Everyone agrees that President Barack Obama is "black" and has some African ancestry. Equating "black" with "African ancestry" seems straightforward in terms of the racial definitions that most people use. However, the situation may not be that simple. It is common knowledge that Barack Obama's mother was white and of European ancestry and his father was black and from Africa. Although one often hears Barack Obama described as "black," he can also be referred to as "half-black." As anthropologist Jon Marks notes in his discussion of another well-known African American, attorney Lani Guinier, we wind up asking how someone can be black and half-black at the same time.[86] The answer is that the two labels refer to different things. When we say that Barack Obama is half-black, this is a statement regarding his immediate genetic ancestry, which can be described as half-white (mother) and half-black (father). When we refer to Barack Obama as black, we are referring to a statement of cultural identity, acknowledging a social group to which he self-identifies.

The contrast between genetic ancestry and cultural identity in this case is clear when we consider the following question (one that I pose to my students): Would you describe Barack Obama as a black man with a white mother or a white man with a black father? Although both are accurate from the viewpoint of biological ancestry, I find that most people pick the first option, an answer that gives us insight into how people view racial categories. As Marks points out, the typical use of race is to classify people using an all-or-nothing classification system—black or white. Historically, this approach can be traced back in the United States to a time when the "one drop of blood" rule was used, stating that anyone with *any* non-white ancestry, even one great-grandparent out of eight, was classified as black.[87] Today, people in the United States self-identify using cultural rules, but only in recent decades have people been able to identify as belonging to two or more races in the census. Someone who has one parent with African ancestry and one parent with European ancestry might self-identify with either or both racial groups.

The problem with equating cultural identity with genetic ancestry is that varying amounts of ancestry are subsumed within a single culturally defined group. Thus, Barack Obama has both black and white ancestry, but is considered black in a cultural sense. Someone else who identifies as black might have a different genetic history. For example, consider the studies cited in the previous myth where some self-identified African Americans actually have more European ancestry than African ancestry. A single cultural category—"black"—can encompass a wide range of genetic histories. There are of course reasons for defining groups by cultural criteria. For example, a person with three grandparents of African ancestry and a person with all four grandparents of African ancestry are both likely to be treated in a similar fashion, including both being labeled as black, even though their ancestral history is different. The problems come when we assume beforehand that these categories imply genetic homogeneity, as when we use race as a diagnostic factor in medicine and public health. Likewise, many have incorrectly assumed that someone's genetic ancestry is an accurate index of their behavior, individually or culturally. Such assumptions have been all too common throughout history leading to countless examples of persecution.

In some cases, cultural identity and genetic ancestry will match up better than others will, but this cannot be determined without analysis. An example that I am familiar with is the case of the Irish Travellers, a nomadic group in Ireland that historically has traveled from town to town looking for seasonal labor, performing odd jobs, and selling goods. In earlier times, the Travellers moved about in horse-drawn caravans, reminiscent to some extent of Roma populations throughout Europe. The cultural similarity of the nomadic lifestyle of the Travellers and

Roma might suggest an ancestral link as well. On the other hand, some have argued that the Travellers cultural similarity to the Roma was coincidental and did not represent a genetic connection. Does cultural similarity necessarily imply an historical (and ancestral) connection? To answer this question, my colleague Michael Crawford of the University of Kansas conducted several studies using red blood cell genetic markers of the Irish Travellers that he collected in 1970. In one study, I worked with Mike to analyze these genetic markers; we found that the Travellers were genetically distinct from Roma populations and most similar to other Irish populations, with a small difference resulting from genetic drift (because the Travellers are a small population).[88] The Irish Travellers are genetically Irish and not Roma. Here, cultural similarity says nothing about genetic ancestry.

Another example of a disconnect between genetic ancestry and cultural identity are the Lemba, an African tribal group in southern Africa.[89] The Lemba have long considered themselves to be Jewish, and are often referred to as "Black Jews." This cultural identity stems from their oral traditions, which state that their male ancestors were Jewish and came from a place named "Sena" to the north and traveled to southern Africa by boat. Although some Lemba now practice Christianity or Islam, some consider themselves Jewish through both ancestry and religious practice. Although the Lemba practice Jewish customs such as certain food taboos and male circumcision, these behaviors are also found in Islam. Some anthropologists did note some aspects of Lemba culture, such as the ritual slaughter of animals, that appeared to have been of Middle Eastern origin and consistent with their oral history of being descended from Jews.[90]

The hypothesis of Jewish ancestry was tested by examining Y chromosome DNA markers in Lemba men because the Y chromosome is passed on from father to son. If the oral traditions of the Lemba are correct and they are descended in part from Jewish males who came to Africa, then we should see some genetic affinity between the Lemba and Jewish populations in the Middle East. Researchers focused on a particular Y chromosome haplotype that tends to be found in highest frequency in Middle Eastern Jewish populations today. They found that 9 percent of the Lemba men had this marker in general, and over 50 percent of men in the Buba clan, the oldest clan and the one that had the strongest oral tradition of Jewish ancestry. However, the mitochondrial DNA (inherited through the female line) of the Lemba was more typically African, consistent with the oral history that male ancestors came from the Middle East.[91] However, more recent analyses of Y chromosome DNA has questioned the specific link with historical Jewish populations.[92]

Even if we assume that further evidence can confirm the initial results of a link to Jewish populations, does this make the Lemba Jewish? No. It only establishes that the Lemba have genetic ancestry from Jewish populations, which does not make them Jewish. Being Jewish is an issue of cultural identity, not genetic ancestry. According to *Halakha* (Jewish religious law), there are two ways someone can be considered a Jew: they either have a Jewish mother or have converted to Judaism. Putting aside conversion for the moment, this means that if a woman is Jewish, then the children are also Jewish. The transmission of cultural identity is through the mother. For example, if a Jewish woman has children with a non-Jewish man, then the children are considered Jewish. However, if a Jewish man has children with a non-Jewish woman, then the children are not considered Jewish, at least by the majority of Jews. From the viewpoint of genetic ancestry, both examples involve one parent who is Jewish and one parent who is not Jewish, but Jewish identity is not determined genetically but culturally, according to the rule of matrilineal descent. Some practitioners of Reform Judaism also practice patrilineal descent under specific conditions, but this is not accepted by many Jews. In the case of the Lemba, their genetic connection to Jewish peoples comes through the paternal line, and therefore they are not Jewish in the eyes of traditional Jewish law (unless they convert).

Having Jewish ancestry does not necessarily make one Jewish, as seen in the Lemba example. On the other hand, being Jewish does not necessarily mean that someone has Jewish genetic ancestry, because of conversion. Of course, there is some overlap between genetic and cultural aspects of being Jewish in that cultural membership is passed on through the mother's line, as are half of one's nuclear DNA. The overlap is partial, but not complete. As a result, it is possible for someone to have Jewish genetic ancestry and be considered Jewish, to have Jewish genetic ancestry and not be considered Jewish (the Lemba), and to have completely different genetic ancestry but be considered Jewish through conversion.

Myth #46 Sickle cell Anemia is a "black disease"

Status: Sickle cell anemia is a genetic disease that reduces the ability or red blood cells to transport oxygen, which can lead to serious medical problems and death. The prevalence of sickle cell anemia in the United States is much higher in African Americans than European Americans, a finding that initially led researchers to suggest that sickle cell anemia was a racial disease—a disease of blacks. We now know that this is a misleading statement. Although some populations in Africa do have elevated frequencies of the sickle cell allele, many do not. Further, a number of populations in the Middle East, Southern Europe, and South Asia also have a higher frequency of this allele. There is no simple association with racial classification when we examine the global pattern of variation.

Sickle cell anemia is a genetic disease that has often been referred to as a "black disease"; that is, found only in people of African ancestry. The actual distribution of sickle cell anemia is more complicated. Sickle cell anemia is a genetic disease that causes distortion of red blood cells, which can lead to a number of medical problems and an increased risk of mortality. Human red blood cells contain the protein hemoglobin, which carries oxygen to tissues and organs. Hemoglobin is made up of different protein chains. One of these, the beta chain, is the focus of this myth. The structure of the beta chain is affected by a gene on our 11th chromosome pair. The most common form of this gene is the hemoglobin *A* allele, which is the normal form. People with two copies of the *A* allele have the genotype *AA*. ("Genotype" refers to the genetic makeup of an individual, defined by what he or she has inherited from both parents; in this case, an *A* allele from both parents.) People with the *AA* genotype have normal hemoglobin. Across our species, the *A* allele is the most common, and many populations have a frequency close to or exactly at 100 percent, which means that most people have normal hemoglobin.

However, the *A* allele is not the only form of the beta hemoglobin gene. As with any other gene or DNA sequence, new forms arise through mutation. There are a number of mutant hemoglobin alleles, including hemoglobin *C*, hemoglobin *E*, and hemoglobin *S*. The *S* allele, the focus of this myth, gets its name because it is linked to the genetic disease known as sickle cell anemia. The beta hemoglobin protein is made up of 146 amino acids (building blocks of proteins), each of which are coded for by a sequence of three DNA bases. The *S* allele is different from the *A* allele because of a single mutation—the sixth DNA letter, A (short for the chemical base adenine), has changed to a T (thymine). The result of this single and simple change is that the amino acid valine is produced instead of the amino acid glutamic acid.

If you have two copies of the *S* allele (the genotype *SS*), you have the genetic disease known as sickle cell anemia. Here, the structure of the red blood cells is changed from their typical donut shape to a sickle shape (hence the name). These deformed blood cells do not transport oxygen as effectively as normal cells and often stick in small blood vessels.

If untreated, sickle cell anemia can lead to many medical problems and death.[93] Individuals that have only one sickle cell allele (and one normal hemoglobin allele) have the genotype *AS*. They are seldom affected and have functional hemoglobin, but they do carry the *S* allele and can therefore pass it on to the next generation. The standard principles of genetic inheritance mean that if two parents each have the *AS* genotype, there is a one in four chance of having a child with sickle cell anemia. (This is analogous to the probability of flipping two coins and getting two tails.)

The first discoveries of sickle cell anemia were made in African Americans. Over time, this association was taken as evidence for a racial trait. In fact, the growing tendency to consider sickle cell disease in racial terms had caused researchers to question the validity of a sickle cell diagnosis in a European American who had sickle cell anemia. Early in the twentieth century the racial association for sickle cell anemia was so powerful that exceptions were considered to be due to African admixture or misdiagnosis. Even after sickle cell anemia was found to occur in other human populations, there remained this tendency to treat sickle cell anemia as a disease of blacks.[94]

It is not too hard to see how this racial association with sickle cell anemia continues. After all, sickle cell anemia is much more common in African Americans that in other ancestral groups. For example, a study of different ethnic groups in California showed that one in 700 blacks had sickle cell anemia, compared with roughly one in 158,000 whites and no cases among those of Asian ancestry.[95] However, this difference does not mean that this association can be equated to a direct relationship between African ancestry and sickle cell anemia. First off, high levels of the sickle cell allele (and thus sickle cell anemia) are not found throughout Africa. Secondly, other populations outside of Africa also have high frequencies of the sickle cell allele. The geographic distribution of the sickle cell allele in our species turns out to reflect natural selection and not race.

Given the increased mortality of individuals with sickle cell anemia, we expect that over time natural selection will act to keep the incidence of the *S* allele close to zero. The frequency of the mutant *S* allele will initially be very low because the probability of an *A* allele mutating into an *S* allele is very low. The rare individual who inherits two copies of the mutant *S* allele will have sickle cell anemia and, until recent medical developments, is likely to die before reaching adulthood. As a result, those individuals with sickle cell anemia will not pass on the *S* allele. Although those who carry the trait (genotype *AS*) can still pass on the *S* allele, the mathematics of natural selection shows that the frequency of *S* would not increase. Even with recurrent mutation, the frequency of

S would be close to or at zero. Indeed, this is the case in large parts of our species where the frequency of S is at very low values between 0 and 1 percent.

However, the frequency of S is higher in other parts of the world, ranging from 1 percent to as high as 20 percent.[96] In Africa, the frequency of S is particularly high in West and Central Africa, but essentially zero in South Africa and North Africa. The fact that African Americans have elevated frequencies of sickle cell anemia is not because it is a disease of blacks, but instead reflects the fact that their African ancestors came from populations primarily in West Africa, a region where the S allele was higher. We do not always see the higher frequencies of S and sickle cell anemia in other parts of Africa. In addition, not all populations with elevated frequencies of the S allele are found in Africa. Higher levels of S have also been found in parts of the Middle East, South India, and Mediterranean Europe. Not all African populations have increased risk for sickle cell anemia, and not all populations that have higher risk for sickle cell anemia are African.

Instead of race or skin color, the key factor affecting the distribution of the sickle cell allele turns out to be a function of geography and ecology. We find elevated frequencies of the S allele (more than 1 percent) in populations that have a history of malaria. Malaria is an infectious disease that is caused by a parasite that infects red blood cells. You cannot transmit malaria to another person by touching or kissing the person—it has to be transmitted by a mosquito bite. When a person is bitten by an infected mosquito, parasites are introduced into the person's blood where they can multiply quickly. Several different species of malarial parasite are spread by different species of mosquito. The most serious form of malaria is caused by the parasite *Plasmodium falciparum*. This type of malaria has long been a serious disease for humanity, with hundreds of millions of people being infected and over a million dying each year.[97]

What does this have to do with the sickle cell allele and sickle cell anemia? The geographic distribution of a history of malarial infection correlates with the frequency of the sickle cell allele. Populations indigenous to regions where there has been frequent malaria have elevated frequencies of the sickle cell allele. It turns out that if you have one sickle cell allele and one normal hemoglobin allele (genotype AS), you are resistant to falciparum malaria. Your blood chemistry is altered enough to provide resistance to malaria, but not so much as to give you sickle cell anemia. In an environment where there is frequent malaria, the AS genotype has the highest probability of survival (fitness). Individuals with two S alleles suffer from sickle cell anemia and have the lowest fitness.

Individuals with normal hemoglobin (*AA*) do not have sickle cell anemia and have higher fitness than those with sickle cell anemia, but are more likely to suffer from malaria, making them less fit than those with one sickle cell allele (*AS*).

The case for sickle cell and malaria represents a classic case of what geneticists call balancing selection (also known as selection for the heterozygote). In terms of genotype, someone can have zero, one, or two copies of the *S* allele. In malarial environments, those individuals with one copy (*AS*) are better off than someone with no copies (*AA*) or someone with two copies (*SS*). We have been able to reconstruct the history of this adaptation in malarial environments. In a non-malarial environment, the sickle cell mutant will be selected against because of the harmful effects of sickle cell anemia, and the frequency will stay low, close to zero. With malaria present, individuals that have one *S* allele will be at an advantage, leading to selection for this allele and an increase in its frequency—the higher the value of the *S* allele, the greater the expected number of individuals with the *AS* genotype who are resistant to malaria. The *S* allele cannot keep increasing, however, because the higher the frequency of *S*, the greater the number of people who will have sickle cell anemia.[98] After a certain point, if *S* increases, the disadvantage of higher values of *S* (sickle cell anemia) will outweigh the advantage (resistance to malaria). Balancing selection leads to a *balance* between the advantage and disadvantage of a mutant allele when those with one copy of the mutant allele have the highest fitness. If the frequency of *S* is too low, then more people are likely to die from malarial infection, but if the frequency is too high, then more people are likely to die from sickle cell anemia. The optimal frequency is the one that produces the fewest overall deaths from both sickle cell anemia and malaria. Over time, natural selection will quickly result in this optimal frequency, which represents a balance of cost and benefit. The balance point today is a frequency of *S* between 1 and 20 percent depending on local history. Today, we find that those populations that have a history of malaria also have elevated frequencies of the sickle cell allele.

We have learned much about the evolution of the sickle cell allele. The late anthropologist Frank Livingstone has argued that the conditions for an increase in malaria in Africa were actually caused by humans.[99] Several thousand years ago, slash-and-burn agriculture spread in West Africa altered the ecology of the area. Slash-and-burn agriculture clears land for farming by burning down trees in the tropical rain forest, which leads to changes in soil chemistry resulting in open pools of standing water, an environment conducive to the spread of mosquitoes. In addition, the growing human population would provide numerous hosts on which the mosquitoes could feed. The net result of these ecological changes is that conditions became hospitable to the growth and spread of the mosquito population. As mosquitoes became more numerous, there would be greater spread of the malarial parasite. Given this change, the sickle cell mutation would have an advantage and increase in frequency until a balance point was reached. Molecular genetic analysis has shown that selection for the sickle cell mutant has actually happened more than once—the evidence suggests several origin points in Africa and another one in India and Saudi Arabia.[100] All of these changes are likely to have occurred only within the past several thousand years, which provides an excellent example that human populations have continued to evolve in recent times (see Myth 48).

A final lesson is about malaria as an agent of human evolution. It is not just the sickle cell allele that has changed in frequency because of selection for malarial resistance. Other genes, such as the Duffy blood group and genes that affect a condition known as thalassemia, have also been affected by adaptation to malaria.[101] Malaria has been a major factor in recent human evolution.

Consideration of the evolution of the sickle cell allele provides nuance and detail about our species' genetic variation that goes beyond a simple racial classification scheme. Although it is true that the rate of sickle cell anemia is much higher in African Americans than in other ethnic groups, this does not make sickle cell anemia a "black disease" in the sense of a strict correlation between ancestry and disease. As shown here, the sickle cell mutations have been selected for in situations where malaria was common, across areas in Africa, Europe, the Middle East, and South Asia. The frequency of the sickle cell allele varies within each of these regions. This is the type of genetic detail that is lost when using broad racial classifications.

Myth 7: There is a strong genetic relationship between brain size and intelligence test scores

Status: Because the fossil record shows an increase in both brain size and cognitive ability during the course of human evolution, we might expect a strong correlation between brain size and cognitive test scores in living humans. Although some studies show a moderate correlation, some of this is due to environmental effects, and the actual genetic correlation is much smaller. We cannot accurately predict test scores from brain size.

It is sometimes assumed that the size of one's head and one's intellectual ability are related. Part of this assumption may derive from a partial understanding of the evolution of human brain size. As noted in previous myths, the increase in brain size over the past several million years is one of the major trends in human evolution, with modern humans having a brain size three times that of *Australopithecus*. The archaeological record also shows a dramatic increase in cognitive ability, including improvements in stone tool manufacture and symbolic behavior. Although brain size is only part of the picture we need to consider the role of structural reorganization of the brain. The fact that brain size does increase so dramatically raises a common question regarding the relationship of brain size and cognitive ability. If later species of *Homo* had larger brains and were capable of more complex behaviors (i.e., were "smarter"), then does this mean that the same relationship holds *within* our species today? That is, are people with larger brains smarter?

Given that "smart" and "intelligence" have different definitions and connotations and may not always be assessed in the same way, let us rephrase that question in terms of something we can and do measure, which is intelligence test scores. Do people with larger brains score higher on intelligence tests? Historically, this question has been hard to answer, and estimates have tended to rely on anecdotal evidence or looking at correlations of external head measures with test scores. Even without any direct evidence relating to brain size and intelligence, it has long been assumed that such a relationship does in fact exist, and that observed differences in brain size between human groups (races) is evidence of differences in intelligence as well.[102]

For a long time, testing the correlation between brain size and intelligence was difficult because brain size could only be measured after someone had died. Today, it is possible to obtain estimates of brain size from living humans using MRI (magnetic resonance imaging) scans, which can then be correlated with the same individual's scores on assorted intelligence tests. Here, I am using correlation in a precise statistical sense as a measure of association between two variables, in this case brain size and intelligence test score. A correlation can range from a value of –1 to +1. A value of +1 is the highest possible positive correlation, meaning that two variables are perfectly correlated. A value of –1 also indicates a perfect relationship, but one that is negative. A correlation of 0 means that there is no relationship whatsoever. Correlation analysis also includes a test of statistical significance, which indicates the probability that an observed correlation could occur by chance.

Several studies have looked at the correlation between brain size (as measured by MRI) and various intelligence tests and the average correlation across a number of studies in in the range of 0.4 to 0.45.[103] This is a moderately strong correlation suggesting a general tendency for individuals with larger brains to score higher on intelligence tests. However, we do not want to read too much into these numbers. For one thing, these results imply only that a portion of variation in test scores is explained by variation in brain size. It turns out (consult any statistics book) that it is easy to figure out how much explained variation we are talking about—you simply square the correlation and multiply by 100. Given correlations of 0.4 to 0.45, this means that between 16 and 20 percent of the variation in test scores can be related to (explained by) variation in brain size. This means that the remainder of the variation in test scores (80 to 84 percent) is not related to variation in brain size.

However, the major problem with interpreting the brain size/test score correlations is that they are *phenotypic* correlations, not *genetic* correlations. The phenotype is what we actually measure, be it brain size or intelligence test score, and it reflects the interaction of both genetic and environmental factors. This means that we cannot necessarily make any inferences about the degree of genetic correlation when we are looking at the phenotypic correlation. Recalling the old adage that correlation does not imply causality, it might be the case that the phenotypic correlation between brain size and test scores reflects other variables that affect both brain size and test scores, making them seem more correlated than would be expected genetically. As an example of how environmental factors can affect the phenotypic correlation, consider socioeconomic status, which is known to have an effect on both intelligence test scores and overall body size (which in turn affects brain size). If there is variation in socioeconomic status in a sample, then we could wind up getting a correlation between brain size and test scores because they are both related to socioeconomic status.[104]

Anthropologist P. Thomas Schoenemann and colleagues developed a way around the problems with looking at phenotypic correlations.[105] They noted that you could get higher phenotypic correlations because of variation between families in environmental factors such as a parent's occupation, a parent's education, and other measures of socioeconomic environment. Instead, they focused their analysis on variation *within families* by looking at brain size and test scores in siblings who share environmental influences. The expectation was that if brain size actually had an effect on test scores, then they should also see a significant correlation in their within-family comparisons. For the most part, they did not. For example, the correlation between brain size and an overall measure of cognitive ability (derived statistically from the results of a number of different intelligence tests given to all subjects) was −0.05, which is not statistically different from zero. If there was a genetic relationship between brain size and test score, then we should see that in paired siblings, the sib with the larger brain size had the higher test score. Sometimes this was the case, but for a number of sib pairs the reverse was true. Overall, there was no correlation between siblings. This zero within-family correlation contrasts strongly with a between-family correlation of 0.45, showing that variation in intelligence test scores had nothing to do with genetic variation and everything to do with environmental factors that varied between families. Although additional studies have shown somewhat higher genetic correlations, they are still less than about 0.2, which means that only a small fraction (4 percent or less) of test score variation is likely to be related to genetic variation in brain size. Schoenemann has concluded that the actual genetic correlation is likely not zero, but is very low, and that it is difficult to see any statistically significant effect unless we had very large samples.[106] These results seem odd because we expect to see a stronger genetic relationship between brain size and cognitive ability given the trends in cranial volume that we see in the fossil record of human evolution. This suggests that the processes of evolution in brain size might be quite different between species than within species.

Although much more research is needed to understand the evolution of human brain size better, it is clear that there is no evidence for a large correlation between brain size and intelligence. The moderate phenotypic correlations appear to reflect environmental differences between families and the underlying genetic correlations are very small.

The assumption that a strong correlation can be used to interpret population differences in intelligence test scores does not hold up. We cannot rank human groups by brain size in an attempt to judge their intellectual prowess. Although there is geographic variation in human brain size, it does not appear to have anything to do with intelligence. Some of the variation in brain size between human populations is related to latitude; we tend to find larger skulls in very cold climates such as the arctic and Siberia because large heads lose heat less rapidly, an adaptation to very cold environments.[107] Apart from these populations, much of the variation in cranial size (and presumably brain size) elsewhere appears to be neutral, reflecting population history but not selection.[108]

Humans are no longer evolving

Status: Major changes in human evolution during the past 12,000 years have been cultural and these changes have continued and accelerated since that time. Because of the rapid nature of cultural change, it is sometimes suggested that we are no longer evolving biologically. Analysis of genetic variation in living humans shows clear examples of recent evolution (during the past 5,000 to 10,000 years), which is likely to continue. Although we may change faster culturally than genetically, we have not stopped evolving genetically. Indeed, our species' cultural evolution has affected our genetic evolution.

Some of the questions I get as an anthropology professor tend to be repeated year after year. One of the most frequent questions is "Are we still evolving?" The short answer is of course "Yes." Evolution is an ongoing process that affects all living creatures. However, the more interesting aspect of this question is to consider why it is asked. Is it because we are so used to considering ourselves as separate from nature? Or, is it because we have increasingly adapted ourselves culturally, such that the major changes we see in humanity are behavioral, nor biological. After all, we are the same as humans two hundred years ago, but during that time, we have seen the development of electricity as an energy source, automobiles, planes, radios, televisions, and computers. During my lifetime so far, I have seen the development of color television, personal computers, space travel, the Internet, and genetic engineering, to name only a few. Yesterday's science fiction has increasingly given rise to today's fact. The changes in our world go beyond technological devices. The human species has increased from less than 2 billion people in 1900 to over 7 billion today, and we continue to develop into a global economy. Not only has the world changed so much, it continues to change at what appears to be an ever-increasing rate. I see this acceleration of cultural change as something that started in the Stone Age but continues today. It may have taken hundreds of thousands of years to go from Acheulean tools to Levallois tools, but only thousands of years to go from small agricultural villages to large civilizations and cities, and only a few years to go from e-mail to smartphones.

However, has cultural change replaced biological evolution, or is it just operating at a much faster pace? Has our species' cultural changes modified our pattern of evolution such that genetic variation, the stuff of evolutionary theory, is no longer relevant? For example, consider the fact that advances in public health and medicine have kept people alive that might have died just a short time ago. If we increasingly buffer ourselves against threats such as infectious disease, does this mean that we have eliminated the possible action of natural selection? Does cultural adaptation allow the relaxation of selective pressures? To some anthropologists and biologists, this has already happened, and evolution has stopped. However, others disagree, pointing to evidence of recent genetic evolution as well as evidence that the process continues today.[109]

The consensus is that *both* biological evolution and cultural change continue to take place, although it is clear that the speed of cultural change is much faster. Further, the rapid changes in human cultural evolution since the origin of agriculture have not supplanted biological change as much as they have changed the environment in which evolution takes place. The underlying conditions for evolution have changed as the human species has increased dramatically in size, dispersed throughout the world, and changed its diets, tools, and structure of societies. In this sense, we can describe human evolution as *biocultural* in nature.

An example that helps show recent human evolution as it relates to cultural change is the case of lactose intolerance. As mammals, human infants rely on mother's milk for nutrition. (Infant formula is a recent invention.) Infants produce the lactase enzyme that allows them to digest lactose (milk sugar). The production of the lactase enzyme is regulated genetically. In other mammalian species, a genetic "switch" shuts down lactase production when the infant is weaned, as it is no longer needed. After nursing, other mammalian species will not drink milk ever again. Most adult mammals are by definition lactose intolerant. The ancestral condition for humans is to be lactose intolerant and anyone who drinks milk or consumes certain dairy products will likely suffer some gastrointestinal discomfort or worse. What makes humans particularly interesting is that not every person in the world is lactose intolerant. A mutant form of the lactase gene leads to lactase persistence, the ability to continue to produce lactose throughout life. The gene has two alleles, one of which is the lactase restriction allele (stopping lactase production after weaning) and the other is the lactase persistence allele (allowing continued production of lactase throughout life). If you possess at least one lactase persistence allele (which is dominant), you are able to drink milk.

Some human populations have very low frequencies of the lactase persistence allele while others have higher frequencies, which means a higher frequency of people that are lactose tolerant (can drink milk) and a lower frequency of people who are lactose intolerant. The correlation of allele frequency and cultural variation is immediately clear—those populations that have a history of dairy farming have a higher frequency of the lactase persistence allele. The lactase persistence allele was due to an initial mutation that was then selected for in populations that practiced dairy farming because it allowed those individuals with a persistence allele to utilize a new form of nutrition.

Dairy farming is a recent human invention, beginning in the Middle East and North Africa between 7,500 and 9,000 years ago and then spreading into Europe, and beginning about 3,300 to 4,500 years ago in sub-Saharan Africa. Molecular analysis has enabled researchers to estimate the age of the initial mutation and the evolution of lactase persistence. It turns out that there were different mutations resulting in lactase persistence in different populations. The specific mutation in Europe dates back between 8,000 and 9,000 years, and the mutation in African populations dates back about 3,000 to 7,000 years ago. These dates correspond to the known history of dairy farming.[110] The key point here is that all of these changes took place in recent times in human evolution and resulted from cultural changes (dairy farming) changing the nature of selection.

Another example is the case of evolution for increased frequency of the sickle cell allele for hemoglobin, which was discussed in Myth 46. Changes in the environment led to increased levels of malaria in some regions, leading to an increase in the frequency of the sickle cell variant because individuals with one copy of that allele are resistant to malaria. Analysis of the molecular genetics of the hemoglobin *S* allele provides an estimate that these changes likely occurred only within the last 2,000 to 3,000 years,[111] a very short time in the evolutionary history of our species. We clearly did *not* stop evolving genetically with the origin of agriculture.

The examples of lactase persistence and hemoglobin *S* are examples of recent evolutionary change in human populations. Given the magnitude of demographic, dietary, and environmental changes due to the origin of agriculture, these types of evolutionary response are not that surprising. As we learn more about human genetic variation, more examples have become known and more will likely be discovered. New methods of examining DNA variation have provided ways of detecting recent natural selection, and many of the DNA regions that show the statistical signatures of selection are those that affect infectious disease resistance and metabolism, both of which have been strongly affected by the change to agriculture.[112] Our cultural evolution (the shift to agriculture) has clearly affected our recent genetic evolution.

There is no argument over the magnitude of cultural changes in the human species in recent evolutionary history or that the rate of cultural change is much faster than the rate of biological evolution. However, it is also clear that biological evolution has *not* stopped for us. Even though cultural changes have buffered humans from certain possible evolutionary fates (such as the protection of antibiotics against many infectious diseases), we have not reached a state where our survival is independent of nature. As the world changes, especially changes due to human existence, new threats will continue to emerge. Global warming, deforestation, and rapid population growth are only a few of the ways in which we alter our relationship with the environment, which in turn can lead to new evolutionary changes. We may have more control over our fate than other species, but we also have a greater impact and do not live apart from nature.

In addition to continued changes due to natural selection, we also continue to evolve due to the action of the other evolutionary forces. Mutation continues all the time to introduce new genetic variants into our gene pool, which can be harmful, helpful, or neutral in effect. In fact, the potential for new mutations is greater today than ever in our species because the number of new mutations is a function of population size and there are more people living today than at any point in the past.[113] More people result in more mutations. Although our overall species is quite large, we are still divided into many smaller local populations, and genetic drift has affected patterns of human variation quite a bit in recent times and will likely continue to do so in the future. Gene flow continues to affect human populations, acting to introduce new alleles into populations and in making populations more similar over time.

We did not stop evolving when we were all hunters and gatherers and we did not stop evolving when we developed agriculture. We did not stop evolving after the Industrial Revolution and we have not stopped evolving today. We continue to change both biologically and culturally. Of course, this leads to another frequent question—where do we go from here? We will return to that question in Myth 50.

Myth #49 Blond hair will eventually disappear

Status: Some have claimed that blond hair will eventually disappear because having blond hair is a recessive genetic trait, and recessive traits will ultimately be replaced by dominant genes. Not only is this a simplistic view of human hair color, but it is based on an inaccurate notion that dominant alleles will replace recessive alleles. In reality, the frequency of an allele does not depend directly on whether it is dominant or recessive, but instead on the evolutionary forces of mutation, selection, drift, and gene flow.

"Blonds 'to die out in 200 years'" was the headline of a story reported by BBC News on the Internet on September 27, 2002.[114] This was one of a number of news items reporting a study that supposedly claimed that blond hair is disappearing from the human species and that the last person with blond hair will be born in Finland about 200 years from now. The story cited two reasons that this would happen. The first reason was that men preferred women who dyed their hair blond rather than have naturally blond hair, implying that over time, natural blond-haired people would become less common. The second reason was that blond hair was a recessive trait, and you would need two copies of the blond gene to have blond hair, under the assumption that recessive traits fade over time.

It was later discovered that the original story that had been picked up by the BBC and other news outlets was a hoax and no such study had been done. Claims that the original study had been conducted by the World Health Organization were also fabricated.[115] It turns out that the idea of the eventual disappearance of blond-haired people has been floating around for some time. The website Snopes.com, which reports on various rumors and urban legends, reports that a similar rash of claims appeared in the media in 1961, some arguing that blond-haired women are more likely to be selected against than brunettes.[116]

Despite the fact that the so-called "study" was in fact a hoax, is there any validity to the claim that blond hair color is slated for extinction? To the best of my knowledge, the suggestion that men prefer artificial blonds to natural blonds and this preference would have an effect on reproductive rates has no support. Further, if we want to argue that natural selection will eliminate blond hair color, then we need to have evidence that hair color plays a significant role in survival and/or reproduction. Lighter shades of hair color are typically found in European populations, and might be related to lighter skin color, as some pigmentation genes have an effect on skin, hair, and eye color. In any event, there does not appear to be any selective disadvantage of lighter hair color in these populations. If we are to believe the myth that blond hair will disappear in the near future, we should be seeing some evidence of this trend and huge differences in survival or reproduction that is linked to hair color. We do not.

There are also problems with the simplistic view of genetics embodied in the myth of the disappearance of blond hair. The misconception starts the idea that the difference between blond and brown hair is due to a single gene with two alleles: a dominant allele, B, which codes for brown hair, and a recessive allele, b, which codes for blond hair. If so, then if you have inherited one or two B alleles from your parents, you will have brown hair, but the only way you can have blond hair is to inherit two copies of the recessive allele. This is an overly simplified view of the genetic of human hair color. In a very general sense, there is evidence that darker shades of hair color tend to be dominant over light shades, but the underlying genetics is considerably more complicated than a single gene with two alleles. Although we are still learning about the specifics, we know that hair color is affected by two different forms of melanin, eumelanin (a black-brown pigment) and pheomelanin (a reddish brown pigment that in high amounts leads to red hair color). Hair color depends on the relative amounts of each type of melanin, where the range from black to blond hair depends on the relative proportion of eumelanin. Hair color is not due to a single gene; there are a number of different genes that have been found to influence pheomelanin production and red hair, and there are at least three different pigmentation genes in Europeans that determine if hair color is brown or blond.[117]

However, the major error in predicting the eventual disappearance of blond-haired people is the view that dominant alleles will replace recessive alleles over time. A related idea is that dominant alleles are more common than recessive alleles. If true, then all dominant alleles will be the most common, and therefore more people will show the dominant trait. However, this is not the case.

A famous example illustrating this error involves a genetic condition known as brachydactyly, where fingers and toes are very short. This condition is due to a dominant allele, which means that if you inherit at least one copy of this allele, you will have this condition. The only way you would have normal-sized fingers and toes is if you inherit two copies of the recessive allele. In 1908, geneticist Reginald Punnett had given a lecture on this topic, after which statistician Udny Yule posed the question of why the frequency of brachydactylic people was not increasing over time, and why there were not many more brachydactylic people in the world. The assumption was that the dominant condition would prevail and increase until there were three brachydactylic people for every person with normal-sized digits. As this is clearly not the case, you can see why Yule was perplexed. Punnett enlisted the help of his mathematician friend, Godfrey Hardy, who quickly resolved the problem, and consequently established an important principle in the further development of evolutionary thought.[118] Whether an allele is dominant or recessive has nothing to do with how common it is in the population. As Punnett and Yule noted, the brachydactyly allele is dominant but is very rare. The frequency of the allele is what matters, not whether it is dominant or recessive. Because the brachydactyly allele is rare, the condition is also rare.

Another example is the ABO blood group, which has three alleles, named *A*, *B*, and *O*, where *O* is recessive and *A* and *B* are codominant (neither is dominant over the other). Your ABO blood type can be type A, B, AB, or O, depending on what alleles you inherit. Because the *O* allele is recessive, the only way you can have type O blood is to inherit two recessive *O* alleles. If we followed the same kind of genetic logic from the myth of disappearing blonds, then we would expect low frequencies of the *O* allele and fewer people with type O blood. The opposite is true. The *O* allele has the highest frequency in most human populations, ranging from slightly less than 50 percent in some isolated groups in Eurasia to 100 percent in some South American native populations.[119] Type O is the most common blood type in the human species and, in some cases (some indigenous South American populations), it is the only blood group present.

Hardy was able to show that the percentage of people who showed different genotypes (combinations of alleles from their parents) was simply a function of the allele frequencies and had nothing to do with whether the allele was dominant or recessive. If the frequency of a recessive allele was high, then there would be many people with the recessive phenotype. Further, Hardy also showed that under certain conditions the allele frequencies would remain the same from generation to generation. Hardy's insights were also independently discovered by physician Wilhelm Weinberg. Consequently, population geneticists now refer to the basic principles relating allele and genotype frequencies as the Hardy–Weinberg law.

Returning to the hair color example, the application of the Hardy–Weinberg law means that under certain conditions the percentage of people with blond hair will remain the same over time. The fact that darker shades tend to be dominant over lighter shades will not change this percentage. Extension of the Hardy–Weinberg law shows that the only way this frequency could change would be the action of one or more of the four evolutionary forces: mutation, selection, genetic drift, and gene flow. Thus, the only way that alleles for blond hair could disappear from our species would be if there was extreme selection against them, and we know of no reason this would be the case. Blonds are not disappearing.

Myth 50 We can predict future human evolution

Status: A common theme in science fiction is the prediction of cultural and biological changes in future human evolution. In reality, it is difficult to make short-term predictions and next to impossible to make any useful long-term predictions about the future of evolution. Some parts of evolution (mutation, drift) are random and their exact future state cannot be predicted in any given situation. When we attempt prediction of human evolution in terms of natural selection, the situation is even more complicated because we would need to be able to predict both cultural and biological futures over many millennia.

In addition to the question of whether we are still evolving (Myth 48), another commonly asked question concerns the future. I am often asked "What do you think humans will look like in X years?" where "X" can range from a thousand or so years to millions and millions of years. If someone asks me what I think humans will look like in a thousand years, my answer is simple. Although allele frequencies are always changing, I expect that we will look pretty much the same as we do today, as a thousand years is too short a span for any significant physical changes to occur (putting aside the more speculative aspects of possible change, such as genetic engineering). If someone asks me what we will look like a million years from now, I have an even simpler answer: I don't know.

The answer "I don't know" is generally disappointing, and some will follow up by pointing out the expectation that science is all about prediction, so a science of human evolution should be capable of producing some specific predictions. We need to be clear here about what is meant by "prediction" as it does not necessarily translate as a forecast of the future. Many sciences are historical by nature, including paleontology, archaeology, and much of astronomy. Here, the focus is not on predicting future events, but instead on using scientific methods to make predictions about what we should see in past events, such as the fossil record.

As an example, consider the hypothesis that the dinosaurs became extinct because an asteroid hit the planet about 65 million years ago. One of the first tests of this hypothesis was based on the prediction that there should be elevated levels of the iridium, an element that is rare on Earth but more common in asteroids and meteors. If the planet was hit by a large object from space 65 million years ago, then we predict that there will be increased levels of iridium in the Earth's crust and mantle that are 65 million years old, which is what we find.[120] Note that this is an example of using prediction for explaining a past event and not for prediction of when the next mass extinction will take place. Many of the myths in this book provide additional examples of how the geological, fossil, and archaeological records have been used to test a variety of hypotheses, such as the idea that all human traits appeared at the same time, the idea that there was only one species of hominin alive at any point in time, and so on. All of these hypotheses (and more) involve using data to test predictions about what we should find in the past and present. Science is always about testing hypotheses, but it is not always about predicting what we will see in the future.

To some extent, we can make some limited predictions about the future course of human evolution, but they consist mostly of demonstrating what could not happen. For example, one of the recurrent themes in science fiction treatments of future human evolution is the idea that our brains will continue to increase in size into the future, resulting in humans with enormous, bulbous heads. This idea focuses on the supposed benefit of larger brains and ignores the cost. Among other problems, any further increase in brain size would require additional prenatal growth, rendering childbirth even more hazardous given the constraints on the pelvic anatomy of a biped. If we consider biological constraints in general, we can rule out a number of other fictional depictions. It is not possible, for example, for humans to evolve to be 7 meters (23 feet) tall or to grow wings and fly.

Apart from setting up some conditions to distinguish between fact and fiction, can we make predictions about the biologically possible alternatives for the course of future evolution? It is difficult to make even short-term predictions in evolution, and it is not possible to make accurate predictions about events many thousands or millions of years into the future. We lack information on all of the possible factors that come into play over time. For example, consider natural selection. It is trivially easy to make predictions about the future state of an allele subject to natural selection, but only if we assume that the current state of the environment and the current values of evolutionary fitness stay the same over time into the far future. This assumption is clearly unreasonable for even a short time as conditions change all the time. We would have to know the future state of all environmental factors, including climate, disease, food resources, predators, and many other factors that could affect natural selection. What is adaptive at one point in time and space may not be in another. Recall the example of the peppered moth (Myth 2). At one point in time, light-colored moths had the advantage, but the environment changed and the advantage shifted to the dark-colored moths (and then back again when the environment changed once more). In order to make any prediction about the future color of the peppered moth, we would need to be able to first predict the condition of the trees (which affects how well the moth is camouflaged), which then requires information about the future state of air pollution (which affects the growth of lichen on the trees). The more detailed we get, the less likely we will have all of the information needed to predict the future course of natural selection.

Even if we did have this information, we would still necessarily have uncertainty about the future because natural selection is not the whole story and there are random elements in evolution—mutation and genetic drift. Although we can make statistical predictions about the average rate of mutation per individual per generation, we do not know beforehand exactly where, when, and what a mutation will be, any more than we can predict beforehand whether a flipped coin will be heads or tails. All we can do is describe relative probabilities—50 percent head and 50 percent tails.

Even in the short run, prediction is problematic when we consider human evolution because our cultural adaptations change the conditions under which biological evolution occurs. As shown in a number of previous myths, the past 12,000 years has seen a series of radical changes resulting from the transition from hunting and gathering to agriculture. The size of the human species has multiplied over a thousand-fold. We live in sedentary populations that are larger than hunting and gathering populations by several orders of magnitude. New diseases appear all the time and old ones evolve to resist the antibiotics we have used in the past. We live in highly stratified societies, often with huge differences in income and health. There are differences in demographic changes, where some populations have the problem of too many people and others have the problem of too few, leading to populations that are aging and shrinking. Whereas the nature of prehistoric societies changed at a slower rate, our societies today are in many ways quite different even from those a generation ago.

We could go on and on, but the point here is that we are living in an ever-changing world and each of the changes listed (and many more) can have an impact on the nature of biological evolution, but it is not clear how this would play out. We would need to know the future cultural status of humanity, which is even more difficult to assess than future biological status, especially over the short term. As I noted above, it is a safe bet that our species will be physically the same over a very short period of 1,000 years because this is too short a time for substantial physical changes of the sort we see over millions of years. In geologic and evolutionary time, a thousand years is the briefest blink of an eye. However, in terms of human cultural change, which has accelerated over time, a thousand years could encompass such change that it might be inconceivable to us today what we will look like in a cultural sense. It is difficult to make social, political, economic, and technological predictions even over the course of a year or two, let alone a thousand years or more. We would not only have to be able to predict a variety of economic, political, and demographic shifts, we would have to be able to predict new technologies and scientific breakthroughs. What would happen, for example, if we discover a way to convert energy into matter or a way to break the limits of the speed of light on interstellar travel? Assuming such things are possible, is there any way we could predict when?

At this point, any predictions we make are not based so much on scientific data but speculations, and these speculations become heavily influenced by whether the person making the prediction is being optimistic or pessimistic.[121] To paraphrase the movie character Dirty Harry, how lucky do we feel? We can roll out a vast array of possible futures, ranging from complete extinction due to global annihilation or catastrophic environmental change, to a future civilization marked by freedom and equality for all spreading out throughout the universe. Use a little imagination and we can come up with a variety of possible futures. The point is that we cannot predict what our cultural or biological futures. From a personal perspective (which I am indulging in for this final myth), I suggest that we prepare for the worst and hope for the best.

We cannot predict the long-term (and sometimes the short-term) future of humanity. Does this mean that the study of evolution has no use? I hope that having come this far in the book that you do not feel this way. Apart from the tangible reward of knowing where we all come from, the study of human evolution also offers us a view of the range of possibilities for our destiny. We have been able to solve a variety of problems to get this far and, purely from a numeric view, we have been successful as a species, certainly more than the many ape species that have died out in the past. We also know from the study of evolution that the scepter of extinction is always above our heads, but what makes us different is our ability to realize this and act upon it. The oft-cited quotation of the philosopher George Santayana applies here: "Those who cannot remember the past are condemned to repeat it."[122]

Notes

1 When modeling human populations, some use a generation length of 20 or 25 years, and others use 30. I use 30 in these examples to match the arguments given in the Rohde *et al.* (2004) paper, although the exact number is not critical for the main points.
2 Livi-Bacci (1997: 31).
3 Weiss (1984).

4 "Match Made in Heaven? Obama is Related to Brad Pitt while Clinton is Cousins with Angelina Jolie." Available at http://www.dailymail.co.uk/home/article-1001741/Match-heaven-Obama-related-Brad-Pitt-Clinton-cousins-Angelina-Jolie.html (accessed July 1, 2014).
5 Rohde et al. (2004). See also the discussion in Olsen's (2003) popular book on genetic history.
6 Feder (2014).
7 Goebel et al. (2008), O'Rourke and Raff (2010), Feder (2014)
8 Goebel et al. (2008).
9 Bradley and Stanford (2004).
10 Goebel et al. (2008), Feder (2014).
11 Bass (1995).
12 Cavalli-Sforza et al. (1994).
13 Goebel et al. (2008), Kemp and Schurr (2010), O'Rourke and Raff (2010).
14 Brown et al. (1998).
15 Fagundes et al. (2008).
16 Bolnick et al. (2012).
17 Goebel et al. (2008).
18 Rasmussen et al. (2014); also see Raff and Bolnick (2014) in the same issue, as well as Balter (2014a).
19 Chatters et al. (2014), see also Balter (2014b) in the same issue.
20 Rasmussen et al. (2015).
21 Raghavan et al. (2014).
22 Heyerdahl (1950).
23 Diamond (1999).
24 Gibbons (2001).
25 Cavalli-Sforza et al. (1994).
26 Diamond (1988).
27 Kayser et al. (2000).
28 Redd et al. (1995).
29 Kayser et al. (2000).
30 Friedlaender et al. (2008).
31 Kayser et al. (2008).
32 Lawler (2010).
33 Diamond (1992).
34 Scarre (2013).
35 Weiss (1984).
36 Larsen (1995, 2003a).
37 Larsen (1995).
38 Cohen (1989).
39 Population Reference Bureau (2015).
40 Cohen (1989).
41 Von Däniken (1970).
42 Feder (2014).

43 Feder (2014).
44 Story (1976), Feder (2014). Both sources give many other examples of flaws in the facts and assumptions of Von Däniken.
45 Lehner (1997).
46 Lehner (1997), Feder (2014).
47 Lehner (1997). See also Lehner's work on building a small pyramid to test different construction methods outlined in the NOVA documentary *This Old Pyramid*.
48 Lewis *et al.* (2010).
49 Hacker (2010).
50 Arias (2014).
51 Hoyert and Xu (2012).
52 Population Reference Bureau (2015).
53 Based on the average of male and female life expectancies for 1900 (47 years) and 2011 (79 years).
54 Armstrong *et al.* (1999).
55 "Leading Causes of Death, 1900–1998," Centers for Disease Control. Available at http://www.cdc.gov/nchs/data/dvs/lead1900_98.pdf (accessed July 11, 2016).
56 Heron (2013).
57 Omran (1971).
58 Centers for Disease Control and Prevention (1999).
59 Omran (1971), Armstrong *et al.* (1999), Centers for Disease Control and Prevention (1999).
60 Heron (2013).
61 "Cause-specific mortality," World Health Organization. Available at http://www.who.int/healthinfo/global_burden_disease/estimates/en/index1.html (accessed July 11, 2016).
62 Levins *et al.* (1994).
63 Armstrong *et al.* (1999).
64 See Relethford (2009) for a similar example.
65 Relethford (2002).
66 Relethford (1997), Jablonski and Chaplin (2000).
67 Robins (1991), Jablonski and Chaplin (2000), Elias and Williams (2013).
68 Osborne and Hames (2014).
69 Jablonski and Chaplin (2000).
70 Elias and Williams (2013).
71 See Long and Kittles (2003) for a comparison of different views of biological race.
72 See Marks (1995) for a discussion of Coon's work and reaction to it.
73 Graves (2001).
74 Relethford (2013b).
75 Darwin (1871: 226.
76 Barbujani and Colonna (2010).

77 Long, Li, and Healy (2009).
78 Relethford (2004).
79 Relethford (2012).
80 The 13 percent estimate is from Tishkoff et al. (2009) and the 19 percent estimate (median) is from Bryc et al. (2010).
81 Parra et al. (1998, 2001).
82 http://www.census.gov/population/socdemo/race/interractab1.txt
83 Parra et al. (2001).
84 Bryc et al. (2010).
85 Tishkoff et al. (2009).
86 Marks (1994).
87 Marks (2011), Chapter 13.
88 Relethford (2013a).
89 Thomas et al. (2000).
90 See "Tudor Parfitt's Remarkable Quest" on the research of Tudor Parfitt. Available at http://www.pbs.org/wgbh/nova/ancient/tudor-parfitts-remarkable-quest.html (accessed July 1, 2014).
91 Thomas et al. (2000).
92 Soodyall (2013).
93 Mielke et al. (2011), Goodman et al. (2012).
94 Goodman et al. (2012).
95 Lorey et al. (1996).
96 Roychoudhury and Nei (1988), Goodman et al. (2012), Jobling et al. (2014).
97 Goodman et al. (2012).
98 This is a well-known principle of population genetics. If the frequency of the S allele is q, then the expected proportion of people with genotype AS is $2q(1-q)$ and the expected proportion of people with genotype SS (and sickle cell anemia) is q^2. See Relethford (2012) for details on the mathematics of balancing selection.
99 Livingstone (1958).
100 Hedrick (2011).
101 Jobling et al. (2014).
102 See Gould (1981) for discussion of the history of measuring brain size.
103 Schoenemann et al. (2000), Schoenemann (2006).
104 Schoenemann (2006).
105 Schoenemann et al. (2000).
106 Schoenemann (2006, 2013).
107 Beals et al. (1984).
108 Relethford (2010).
109 See Balter (2005) for a good review of this debate.
110 Tishkoff et al. (2007).
111 Reviewed in Mielke, Konigsberg, and Relethford (2011: ch. 7).
112 Hawks et al. (2007), Akey (2010).

113 Hawks *et al.* (2007).
114 "Blondes 'to die out in 200 years.'" Available at http://news.bbc.co.uk/2/hi/health/2284783.stm (accessed July 1, 2014).
115 Balter (2005).
116 "Gone Blonde: Are Natural Blondes likely to be Extinct within 200 Years?" Available at http://www.snopes.com/science/stats/blondes.asp (accessed July 1, 2014).
117 Mielke *et al.* (2011: ch. 12).
118 Relethford (2012: ch. 2).
119 Roychoudhury and Nei (1988).
120 Alvarez *et al.* (1980).
121 Relethford (2013b: epilogue.
122 "George Santayana." Available at http://en.wikipedia.org/wiki/George_Santayana (accessed July 11, 2016).

REFERENCES

Aiello, L.C. (2010) Five years of *Homo floresiensis*. *American Journal of Physical Anthropology* **142**, 167–179.

Aiello, L.C. and Wheeler, P. (1995) The expensive-tissue hypothesis: The brain and the digestive system in human and primate evolution. *Current Anthropology* **36**, 199–221.

Akey, J.M. (2010) Constructing genomic maps of positive selection in humans: Where do we go from here? *Genome Research* **19**, 711–722.

Alvarez, L.W., Alvarez, W., Asaro, F., and Michel, H.V. (1980) Extraterrestrial cause for the Cretaceous-Tertiary extinction. *Science* **208**, 1095–1108.

Ardrey, R. (1961) *African Genesis*. Dell, New York.

Arias, E. (2014) United States life tables, 2010. National Vital Statistics Reports, vol. 63, no. 7. National Center for Health Statistics, Hyattsville, MD.

Armstrong, E. (1983) Relative brain size and metabolism in mammals. *Science* **220**, 1302–1304.

Armstrong, G.L., Conn, L.A., and Pinner, R.W. (1999) Trends in infectious disease mortality in the United States during the 20th century. *JAMA* **281**, 61–66.

Arsuaga, J.L., Martinez, I., Arnold, L.J., et al. (2014) Neandertal roots: Cranial and chronological evidence from Sima de los Huesos. *Science* **344**, 1358–1363.

Asfaw, B., White, T., Lovejoy, O., et al. (1999) *Australopithecus garhi*: A new species of early hominid from Ethiopia. *Science* **284**, 629–635.

Baab, K.L., McNulty, K.P., and Harvati, K. (2013) *Homo floresiensis* contextualized: A geometric morphometric comparative analysis of fossil and pathological human samples. *PLoS One* 8(7), e69119. doi:10.1371/journal.pone.0069119.

Balter, M. (2005) Are humans still evolving? *Science* **309**, 234–237.

Balter, M. (2014a) Ancient infant was ancestor of today's Native Americans. *Science* **343**, 716–717.

50 Great Myths of Human Evolution: Understanding Misconceptions about Our Origins, First Edition. John H. Relethford.
© 2017 John Wiley & Sons, Inc. Published 2017 by John Wiley & Sons, Inc.

Balter, M. (2014b) Bones from a watery "black hole" confirm first American origins. *Science* **344**, 680-681.

Barbujani, G. and Colonna, V. (2010) Human genetic diversity: Frequently asked questions. *Trends in Genetics* **26**, 285–295.

Bass, W.M. (1995) *Human Osteology: A Laboratory and Field Manual*, 4th edn. Missouri Archaeological Society, Columbia, MO.

Beals, K.L., Smith, C.L., and Dodd, S.M. (1984) Brain size, cranial morphology, climate, and time machines. *Current Anthropology* **25**, 301–330.

Benton, M.J. (2003) *When Life Nearly Died: The Greatest Mass Extinction of All Time*. Thames & Hudson, London.

Berger, L.R. (2013) The mosaic nature of *Australopithecus sediba*. *Science* **340**, 163.

Berger, L.R., Hawks, J., De Ruiter, D.J., *et al.* (2015) *Homo nadeli*, a new species of the genus *Homo* from the Dinaledi Chamber, South Africa. *eLife* **4**, e09560. doi:10.7554/eLife.09560.

Berger, L.R. and McGraw, W.S. (2007) Further evidence for eagle predation of, and feeding damage on, the Taung child. *South African Journal of Science* **103**, 496–498.

Berger, T.D. and Trinkaus, E. (1995) Patterns of trauma among the Neandertals. *Journal of Archaeological Science* **22**, 841–852.

Bermúdez de Castro, J.M., Carbonell, E., Cáceres, I., *et al.* (1999) The TD6 (Aurora stratum) hominid site: Final remarks and new questions. *Journal of Human Evolution* **37**, 695–700.

Berna, F., Goldberg, P., Horwitz, L.K., *et al.* (2012) Microstratigraphic evidence of in situ fire in the Acheulean strata of Wonderwerk Cave, Northern Cape province, South Africa. *Proceedings of the National Academy of Sciences* **109**, E1215–E1220.

Bishop, G.F., Thomas, R.K., Wood, J.A., and Gwon, M. (2010) American's scientific knowledge and beliefs about human evolution in the year of Darwin. *Reports of the National Center for Science Education*, **30**, 16–18. Available online at http://ncse.com/rncse/30/3/americans-scientific-knowledge-beliefs-human-evolution-year (accessed July 12, 2016).

Boaz, N.T. and Ciochon, R.L. (2004) *Dragon Bone Hill: An Ice-Age Saga of Homo erectus*. Oxford University Press, New York.

Bolnick, D.A., Feder, K.L., Lepper, B.T., and Barnhart, T.A. (2012) Civilizations lost and found: Fabricating history. Part three: Real messages in DNA. *Skeptical Inquirer* **36**, 48–51.

Bradley, B. and Stanford, D. (2004) The North Atlantic ice-edge corridor: A possible Palaeolithic route to the New World. *World Archaeology* **36**, 459–478.

Bräuer, G. (1984) The "Afro-European *sapiens*-hypothesis" and hominid evolution in East Asia during the Late Middle and Upper Pleistocene. *Courier Forschungsinstitut Senckenberg* **69**, 145–165. (Reprinted in Ciochon, R.L. and Fleagle, J.G. (eds) (1993) *The Human Evolution Source Book*. Prentice Hall, Englewood Cliffs, NJ, pp. 446–460.)

Brown, M.D., Hosseini, S.H., Torroni, A., *et al.* (1998) mtDNA haplogroup X: An ancient link between Europe/Western Asia and North America? *American Journal of Human Genetics* **63**, 1852–1861.

Brown, P., Sutikna, T., Morwood, M.J., *et al.* (2004) A new small-bodied hominin from the Late Pleistocene of Flores, Indonesia. *Nature* **431**, 1055–1061.

Brues, A.M. (1977) *People and Races*. Macmillan, New York.

Bryc, K., Auton, A., Nelson, M.R., *et al.* (2010) Genome-wide patterns of population structure and admixture in West Africans and African Americans. *Proceedings of the National Academy of Sciences* **107**, 786–791.

Bustamante, C.D. and Henn, B.M. (2010) Shadows of early migrations. *Nature* **468**, 1044–1045.

Campbell, B.G., Loy, J.D., and Cruz-Uribe, K. (2006) *Humankind Emerging*, 9th edn. Pearson, Boston.

Cann, R.L., Stoneking, M., and Wilson, A.C. (1987) Mitochondrial DNA and human evolution. *Nature* **325**, 31–36.

Cartmill, M. (2008) Review of *Bigfoot Exposed: An Anthropologist Examines America's Enduring Legend* and *Sasquatch: Legend Meets Science*. *American Journal of Physical Anthropology* **135**, 117–120.

Cartmill, M. and Smith, F.H. (2009) *The Human Lineage*. Wiley-Blackwell, Hoboken, NJ.

Caspari, R. (2011) The evolution of grandparents. *Scientific American* **305**(2), 44–49.

Cavalli-Sforza, L.L., Menozzi, P., and Piazza, A. (1994) *The History and Geography of Human Genes*. Princeton University Press, Princeton, NJ.

Centers for Disease Control and Prevention (1999) Achievements in public health, 1900–1999: Control of infectious disease. *Morbidity and Mortality Weekly Report* **48**(29), 621–629.

Chatters, J.C., Kennett, D.J., Asmerom, Y., *et al.* (2014) Late Pleistocene human skeleton and mtDNA link Paleoamericans and modern Native Americans. *Science* **344**, 750–754.

Ciochon, R., Olsen, J., and James, J. (1990) *Other Origins: The Search for the Giant Ape in Human Prehistory*. Bantam Books, New York.

Cohen, M.N. (1989) *Health and the Rise of Civilization*. Yale University Press, New Haven, CT.

Conroy, G.C. and Pontzer, H. (2012) *Reconstructing Human Origins: A Modern Synthesis*, 3rd edn. W.W. Norton, New York.

Conroy, G.C., Weber, G.W., Seidler, H., *et al.* (2000) Endocranial capacity of the Bodo cranium determined from three-dimensional computed tomography. *American Journal of Physical Anthropology* **113**, 111–118.

Cook, L.M., Dennis, R.L.H., and Mani, G.H. (1999) Melanic morph frequency in the peppered moth in the Manchester area. *Proceedings of the Royal Society of London B* **266**, 293–297.

Daegling, D.J. (2004) *Bigfoot Exposed: An Anthropologist Examines America's Enduring Legend*. AltaMira Press, Walnut Creek, CA.

Dart, R.A. and Craig, D. (1959) *Adventures with the Missing Link*. Viking, New York.

Darwin, C. (1859) *On The Origin of Species, by Means of Natural Selection, or the Preservation of Favoured Races in the Struggle for Life*. John Murray, London.

Darwin, C. (1871) *The Descent of Man, and Selection in Relation to Sex*. John Murray, London.

DeGusta, D., Gilbert, W.H., and Turner, S.P. (1999) Hypoglossal canal size and hominid speech. *Proceedings of the National Academy of Sciences* **96**, 1800–1804.

De Heinzelin, J., Clark, J.D., White, T., et al. (1999) Environment and behavior of 2.5-million-year-old Bouri hominids. *Science* **284**, 625–629.

Diamond, J. (1988) Express train to Polynesia. *Nature* **336**, 307–308.

Diamond, J. (1992) *The Third Chimpanzee: The Evolution and Future of the Human Animal*. HarperCollins, New York.

Diamond, J. (1999) *Guns, Germs, and Steel: The Fate of Human Societies*. W.W. Norton, New York.

Díez, J.C., Fernández-Jalvo, Y., Rosell, J., and Cácares, I. (1999) Zooarchaeology and taphonomy of Aurora Stratum (Gran Dolina, Sierra de Atapuerca, Spain). *Journal of Human Evolution* **37**, 623–652.

Disotell, T.R. (2013) Genetic perspectives on ape and human evolution. In Begun, D.R. (ed.) *A Companion to Paleoanthropology*. Wiley-Blackwell, Chichester, UK, pp. 291–305.

Duarte, C., Maurício, J., Pettitt, P.B., et al. (1999) The early Upper Paleolithic human skeleton from the Abrigo do Lagar Velho (Portugal) and modern human emergence in Iberia. *Proceedings of the National Academy of Sciences* **96**, 7604–7609.

Elias, P.M. and Williams, M.L. (2013) Re-appraisal of current theories for the development and loss of epidermal pigmentation in hominins and modern humans. *Journal of Human Evolution* **64**, 687–692.

Eller, E., Hawks, J., and Relethford, J.H. (2004) Local extinction and recolonization, species effective population size, and modern human origins. *Human Biology* **76**, 689–709.

Fagundes, N.J.R., Kanitz, R., Eckert, R., et al. (2008) Mitochondrial population genomics supports a single pre-Clovis origin with a coastal route for the peopling of the Americas. *American Journal of Human Genetics* **82**, 583–592.

Falk, D. (1992) *Braindance: New Discoveries about Human Origins and Brain Evolution*. Henry Holt, New York.

Falk, D. (2000) *Primate Diversity*. W.W. Norton, New York.

Falk, D., Hildebolt, C., Smith, K., et al. (2009) LB1's virtual endocast, microcephaly, and hominin brain evolution. *Journal of Human Evolution* **57**, 597–607.

Feder, K.L. (2014) *Frauds, Myths, and Mysteries: Science and Pseudoscience in Archaeology*, 8th edn. McGraw-Hill, New York.

Friedlaender, J.S., Friedlaender, F.R., Reed, F.A., et al. (2008) The genetic structure of Pacific Islanders. *PLoS Genetics* **4**(1): e19. doi:10.1371/journal.pgen.0040019.

Gibbons, A. (2001) The peopling of the Pacific. *Science* **291**, 1735–1737.

Gibbons, A. (2006) *The First Human: The Race to Discover our Earliest Ancestors*. Doubleday, New York.
Gibbons, A. (2009) A new kind of ancestor: *Ardipithecus* unveiled. *Science* 326, 36–40.
Gibbons, A. (2010) Close encounters of the prehistoric kind. *Science* 328, 680–684.
Gibbons, A. (2012) Turning back the clock: Slowing the pace of prehistory. *Science* 338, 189–191.
Goebel, T., Waters, M.R., and O'Rourke, D.H. (2008) The Late Pleistocene dispersal of modern humans in the Americas. *Science* 319, 1497–1502.
Goodall, J. (1971) *In the Shadow of Man*. Houghton-Mifflin, Boston.
Goodman, A.H., Moses, Y.T., and Jones, J.L. (2012) *Race: Are We So Different?* Wiley-Blackwell, Chichester, UK.
Goren-Inbar, N., Alperson, N., Kislev, M.E., et al. (2004) Evidence of hominin control of fire at Gesher Benot Ya'aqov, Israel. *Science* 304, 725–727.
Gould, S.J. (1977) *Ever Since Darwin: Reflections in Natural History*. W.W. Norton, New York.
Gould, S.J. (1981) *The Mismeasure of Man*. W.W. Norton, New York.
Gramling, C. (2016) The "hobbit" was a separate species of human, new dating reveals. *Science*, March 30. Available online at http://www.sciencemag.org/news/2016/03/hobbit-was-separate-species-human-new-dating-reveals (accessed July 12, 2016).
Graves, J.L. Sr (2001) *The Emperor's New Clothes: Biological Theories of Race at the Millennium*. Rutgers University Press, New Brunswick, NJ.
Green, R.E., Krause, J., Briggs, A.W., et al. (2010) A draft sequence of the Neandertal genome. *Science* 328, 710–722.
Gunz, P., Neubauer, S., Maurelle, B., and Hublin, J.-J. (2010) Brain development after birth differs between Neanderthals and modern humans. *Current Biology* 20, R921–R922.
Hacker, J.D. (2010) Decennial life tables for the white population of the United States, 1790–1900. *Historical Methods* 43, 45–79.
Hailer, F., Kutschera, V.E., Hallström, B.M., et al. (2012) Nuclear genomic sequences reveal that polar bears are an old and distinct bear lineage. *Science* 336, 344–347.
Haile-Selassie, Y., Gibert, L., Mellilo, S.M., et al. (2015) New species from Ethiopia further expands Middle Pliocene hominin diversity. *Nature* 521, 483–488.
Haile-Selassie, Y., Suwa, G., and White, T.D. (2004) Late Miocene teeth from Middle Awash, Ethiopia, and early hominid dental evolution. *Science* 303, 1503–1505.
Hammond, A.S. and Ward, C.V. (2013) *Australopithecus* and *Kenyanthropus*. In Begun, D.R. (ed.) *A Companion to Paleoanthropology*. Wiley-Blackwell, Chichester, UK, pp. 434–456.
Harmand, S., Lewis, J.E., Feibel, C.S., et al. (2015) 3.3-million-year-old stone tools from Lomekwi 3, West Turkana, Kenya. *Nature* 521, 310–315.

Harvati-Papatheodorou, K. (2013) Neanderthals. In Begun, D.R. (ed.) *A Companion to Paleoanthropology*. Wiley-Blackwell, Chichester, UK, pp. 538–556.

Hawks, J., Wang, E.T., Cochran, G.M., *et al.* (2007) Recent acceleration of human adaptive evolution. *Proceedings of the National Academy of Sciences* **104**, 20753–20758.

Hedrick, P.W. (2011) Population genetics of malaria resistance in humans. *Heredity* **107**, 283–304.

Henneberg, M., Eckhardt, R.B., Chavanaves, S., and Hsü, K.J. (2014) Evolved developmental homeostasis disturbed in LB1 from Flores, Indonesia, denotes Down syndrome and not diagnostic traits of the invalid species *Homo floresiensis*. *Proceedings of the National Academy of Sciences* **111**, 11967–11972.

Heron, M. (2013) Deaths: Leading causes for 2010. *National Vital Statistics Reports*, vol. **62**, no. 6. National Center for Health Statistics, Hyattsville, MD.

Hershkovitz, I., Marder, O., Ayalon, A., *et al.* (2015) Levantine cranium from Manot Cave (Israel) foreshadows the first European modern humans. *Nature* **520**, 216–219.

Heyerdahl, T. (1950) *Kon-Tiki: Across the Pacific by Raft*. Rand McNally, New York.

Higham, T., Jacobi, R., Julien, M., *et al.* (2010) Chronology of the Grotte du Renne (France) and implications for the context of ornaments and human remains within the Châtelperronian. *Proceedings of the National Academy of Sciences* **107**, 20234–20239.

Hodgson, J.A. and Disotell, T.R. (2008) No evidence of a Neanderthal contribution to modern human diversity. *Genome Biology* **9**, 206. doi:10.1186/gb-2008-9-2-206.

Hoffman, A. (2016) Did modern humans wipe out the "Hobbits"? *National Geographic*, March 30. Available online at http://news.nationalgeographic.com/2016/03/160330-hobbits-humans-flores-cave-older-species/ (accessed July 12, 2016).

Holloway, R.L., Broadfield, D.C., and Yuan, M.S. (2004) *Brain Endocasts: The Paleoneurological Evidence. Volume 3, The Human Fossil Record* (ed. Schwartz, J.H. and Tattersall, I.). Wiley-Liss, New York.

Hooton, E.A. (1947) *Up From the Ape*, rev. edn. Macmillan, New York.

Howell, F.C. (1957) The evolutionary significance of variation and varieties of "Neanderthal" man. *Quarterly Review of Biology* **32**, 330–347.

Hoyert, D.L. and Xu, J. (2012) Deaths: Preliminary data for 2011. *National Vital Statistics Reports*, vol. **61**, no. 6. National Center for Health Statistics, Hyattsville, MD.

Indriati, E., Swisher, C.C., III, Lepre, C., *et al.* (2011) The age of the 20 meter Solo River Terrace, Java, Indonesia and the survival of *Homo erectus* in Asia. *PLoS One* **6**(6), e21562. doi:10.1371/journal.pone.0021562.

Jablonski, N.G. and Chaplin, G. (2000) The evolution of human skin coloration. *Journal of Human Evolution* **39**, 57–106.

Jobling, M., Hollox, E., Hurles, M., *et al.* (2014) *Human Evolutionary Genetics*, 2nd edn. Garland Science, New York.

Johanson, D. and Edey, M. (1981) *Lucy: The Beginnings of Humankind*. Simon & Schuster, New York.

Johanson, D. and Edgar, B. (2006) *From Lucy to Language*, rev., updated, and exp. edn. Simon & Schuster, New York.

Jungers, W.L., Larson, S.G., Harcourt-Smith, W., et al. (2009) Descriptions of the lower limb skeleton of *Homo floresiensis*. *Journal of Human Evolution* **57**, 538–554.

Kayser, M., Brauer, S., Weiss, S., et al. (2000) Melanesian origin of Polynesian Y chromosomes. *Current Biology* **10**, 1237–1246.

Kayser, M., Lao, O., Saar, K., et al. (2008) Genome-wide analysis indicates more Asian than Melanesian ancestry of Polynesians. *American Journal of Human Genetics* **82**, 194–198.

Kemp, B.M. and Schurr, T.G. (2010) Ancient and modern genetic variation in the Americas. In Auerbach, B.M. (ed.) *Human Variation in the Americas*. Center for Archaeological Investigations, Southern Illinois University, Carbondale, pp. 12–50.

Kennedy, K.A.R. (1976) *Human Variation in Time and Space*. W.C. Brown, Dubuque, IA.

Kennedy, K.A.R. (2000) *God-Apes and Fossil Men: Paleoanthropology of South Asia*. University of Michigan Press, Ann Arbor.

King, W. (1864) The reputed fossil man of the Neanderthal. *The Quarterly Journal of Science* **1**, 88–97.

Klein, R.G. (2009) *The Human Career: Human Biological and Cultural Origins*, 3rd edn. University of Chicago Press, Chicago.

Krantz, G.S. (1986) A species named from footprints. *Northwest Anthropological Research Notes* **19**, 93–99.

Krause, J., Lalueza-Fox, C., Orlando, L., et al. (2007) The derived *FOXP2* variant of modern humans was shared with Neandertals. *Current Biology* **17**, 1908–1912.

Krings, M., Stone, A., Schmitz, R.W., et al. (1997) Neandertal DNA sequences and the origin of modern humans. *Cell* **90**, 19–30.

Krogman, W.M. (1951) The scars of human evolution. *Scientific American* **185**(6), 54–57.

Kubo, D., Kono, R.T., and Kaifu, Y. (2013) Brain size of *Homo floresiensis* and its evolutionary implications. *Proceedings of the Royal Society B: Biological Sciences* **280**, 20130338. doi:10.1098/rspb.2013.0338.

Lahr, M.M. (1996) *The Evolution of Modern Human Diversity: A Study of Cranial Variation*. Cambridge University Press, Cambridge.

Lalueza-Fox, C. and Gilbert, M.T.P. (2011) Paleogenomics of archaic humans. *Current Biology* **21**, R1002–R1009. doi: 10.1016/j.cub.2011.11.021.

Langergraber, K.E., Prüfer, K., Rowney, C., et al. (2012) Generation times in wild chimpanzees and gorillas suggest earlier divergence times in great ape and human evolution. *Proceedings of the National Academy of Sciences* **109**, 15716–15721.

Larsen, C.S. (1995) Biological changes in human populations with agriculture. *Annual Review of Anthropology* **24**, 185–213.

Larsen, C.S. (2003a) Animal source foods and human health during evolution. *The Journal of Nutrition* **133**, 3893S–3897S.

Larsen, C.S. (2003b) Equality for the sexes in human evolution? Early hominid sexual dimorphism and implications for mating systems and social behavior. *Proceedings of the National Academy of Sciences* **100**, 9103–9104.

Lawler, A. (2010) Beyond *Kon-Tiki*: Did Polynesians sail to South America? *Science* **328**, 1344–1347.

Leakey, M.D. (1979) *Olduvai Gorge: My Search for Early Man*. Collins, London.

Lehner, M. (1997) *The Complete Pyramids: Solving the Ancient Mysteries*. Thames and Hudson, London.

Leonard, W.R. and Robertson, M.L. (1995) Energetic efficiency of human bipedality. *American Journal of Physical Anthropology* **97**, 335–338.

Lepre, C.J., Roche, H., Kent, D.V., et al. (2011) An earlier origin for the Acheulian. *Nature* **477**, 82–85.

Levins, R., Awerbuch, T., Brinkmann, U., et al. (1994) The emergence of new diseases. *American Scientist* **82**, 52–60.

Lewin, R. (1987) The unmasking of mitochondrial Eve. *Science* **238**, 24–26.

Lewin, R. (1997) *Bones of Contention: Controversies in the Search for Human Origins*, 2nd edn. University of Chicago Press, Chicago.

Lewis, B., Jurmain, R., and Kilgore, L. (2010) *Understanding Humans: Introduction to Physical Anthropology and Archaeology*. 10th edn. Wadsworth, Belmont, CA.

Lieberman, D.E. (2001) Another face in our family tree. *Nature* **410**, 419–420.

Lieberman, D.E. and McCarthy, R.C. (1999) The ontogeny of cranial base angulation in humans and chimpanzees and its implications for reconstructing pharyngeal dimensions. *Journal of Human Evolution* **36**, 487–517.

Lieberman, P. and Crelin, E.S. (1971) On the speech of Neanderthal Man. *Linguistic Inquiry* **2**, 203–222.

Lieberman, P., Crelin, E.S., and Klatt, D.H. (1972) Phonetic ability and related anatomy of the newborn and adult human, Neanderthal Man, and the chimpanzee. *American Anthropologist* **74**, 287–307.

Livi-Bacci, M. (1997) *A Concise History of World Population*. 2nd edn. Blackwell, Malden, MA.

Livingstone, F.B. (1958) Anthropological implications of sickle cell gene distribution in West Africa. *American Anthropologist* **60**, 533–562.

Long, J.C. and Kittles, R.A. (2003) Human genetic diversity and the nonexistence of biological races. *Human Biology* **75**, 449–471.

Long, J.C., Li, J., and Healy, M.E. (2009) Human DNA sequences: More variation and less race. *American Journal of Physical Anthropology* **139**, 23–34.

Lorey, F.W., Arnopp, J., and Cunningham, G.C. (1996) Distribution of hemoglobinopathy variants by ethnicity in a multiethnic state. *Genetic Epidemiology* **13**, 501–512.

Louchart, A., Wesselman, H., Blumenschine, R.J., et al. (2009) Taphonomic, avian, and small-vertebrate indicators of *Australopithecus ramidus* habitat. *Science* **326**, 66e1–66e4.

Lovejoy, C.O. (1980) The origin of man. *Science* **211**, 341–350.

Lovejoy, C.O., Suwa, G., Simpson, S.W., et al. (2009) The great divides: *Ardipithecus ramidus* reveals the postcrania of our last common ancestors with African apes. *Science* **326**, 100–106.

Lovejoy, C.O., Suwa, G., Spurlock, L., et al. (2009) The pelvis and femur of *Ardipithecus ramidus*: The emergence of upright walking. *Science* **326**, 71e1–71e6.

Loxton, D. and Prothero, D.R. (2013) *Abominable Science! Origins of the Yeti, Nessie, and Other Famous Cryptids*. Columbia University Press, New York.

Marks, J. (1994) Black, white, other. *Natural History* **103**, 32–35.

Marks, J. (1995) *Human Biodiversity: Genes, Race, and History*. Aldine de Gruyter, New York.

Marks, J. (2011) *The Alternative Introduction to Biological Anthropology*. Oxford University Press, New York.

Martínez, I., Arsuaga, J.L., Quam, R.D., et al. (2008) Human hyoid bones from the middle Pleistocene site of the Sima de los Huesos (Sierra de Atapuerca, Spain). *Journal of Human Evolution* **54**, 118–124.

Martínez, I., Rosa, M., Arsuaga, J.L., et al. (2004) Auditory capacities in Middle Pleistocene humans from the Sierra de Atapuerca in Spain. *Proceedings of the National Academy of Sciences* **101**, 9976–9981.

McDougall, I., Brown, F.H., and Fleagle, J.G. (2005) Stratigraphic placement and age of modern humans from Kibish, Ethiopia. *Nature* **433**, 733–736.

Meikle, W.E. and Parker, S.T. (1994) *Naming Our Ancestors: An Anthology of Hominid Taxonomy*. Waveland, Prospect Heights, IL.

Meldrum, J. (2006) *Sasquatch: Legend Meets Science*. Tom Doherty, New York.

Mielke, J.H., Konigsberg, L.W., and Relethford, J.H. (2011) *Human Biological Variation*, 2nd edn. Oxford University Press, New York.

Milne, D.H. and Schafersman, S.D. (1983) Dinosaur tracks, erosion marks, and midnight chisel work (but no human footprints) in the Cretaceous limestone of the Paluxy River Bed, Texas. *Journal of Geological Education* **31**, 111–123.

Morwood, M. and van Oosterzee, P. (2007) *A New Human: The Startling Discovery and Strange Story of the "Hobbits" of Flores, Indonesia*. HarperCollins, New York.

Olsen, S. (2003) *Mapping Human History: Genes, Race, and Our Common Origins*. Houghton-Mifflin, Boston.

Omran, A. (1971) The epidemiologic transition: A theory of the epidemiology of population change. *The Milbank Memorial Fund Quarterly* **49**, 509–538.

Orlando, L., Ginolhac, A., Zhang, G., et al. (2013) Recalibrating *Equus* evolution using the genome sequence of an early Middle Pleistocene horse. *Nature* **499**, 74–78.

O'Rourke, D.G. and Raff, J.A. (2010) The human genetic history of the Americas: The final frontier. *Current Biology* **20**, R202–R207.

Orr, C.M., Tocheri, M.W., Burnett, S.E., et al. (2013) New wrist bones of *Homo floresiensis* from Liang Bua (Flores, Indonesia). *Journal of Human Evolution* **64**, 109–129.

Osborne, D.L. and Hames, R. (2014) A life history perspective on skin cancer and the evolution of skin pigmentation. *American Journal of Physical Anthropology* **153**, 1–8.

Parra, E.J., Kittles, R.A., Argyropoulos, G., et al. (2001) Ancestral proportions and admixture dynamics in geographically defined African Americans living in South Carolina. *American Journal of Physical Anthropology* **114**, 18–29.

Parra, E.J., Marchini, A., Akey, J., et al. (1998) Estimating African American admixture proportions by use of population-specific alleles. *American Journal of Human Genetics* **63**, 1839–1851.

Pennisi, E. (2013) More genomes from Denisova Cave show mixing of early human groups. *Science* **340**, 799.

Pilbeam, D. (1972) *The Ascent of Man*. Macmillan, New York.

Population Reference Bureau (2015) 2015 World population data sheet. Population Reference Bureau, Washington, DC. Available online at http://www.prb.org/pdf15/2015-world-population-data-sheet_eng.pdf (accessed July 13, 2016).

Potts, R. (1984) Home bases and early hominids. *American Scientist* **72**, 338–347.

Pruetz, J.D. and Bertolani, O. (2007) Savanna chimpanzees, *Pan troglodytes verus*, hunt with tools. *Current Biology* **17**, 412–417.

Raff, J.A. and Bolnick, D.A. (2014) Genetic roots of the first Americans. *Nature* **506**, 162–163.

Raghavan, M., Skoglund, P., Graf, K.E., et al. (2014) Upper Palaeolithic Siberian genome reveals dual ancestry of Native Americans. *Nature* **505**, 87–91.

Rasmussen, M., Anzick, S.L., Waters, M.R., et al. (2014) The genome of a Late Pleistocene human from a Clovis burial site in western Montana. *Nature* **506**, 225–229.

Rasmussen, M., Sikora, M., Albrechtsen, A., et al. (2015) The ancestry and affiliations of Kennewick Man. *Nature* **523**, 455–458.

Raup, D.M. (1991) *Extinction: Bad Genes or Bad Luck?* W.W. Norton, New York.

Ravosa, M.J. (1988) Browridge development in Cercopithecidae: A test of two models. *American Journal of Physical Anthropology* **76**, 535–555.

Reader, J. (2011) *Missing Links: In Search of Human Origins*. Oxford University Press, New York.

Redd, A.J., Takezaki, N., Sherry, S.T., et al. (1995) Evolutionary history of the COII/tRNA^{Lys} intergenic 9 base pair deletion in human mitochondrial DNAs from the Pacific. *Molecular Biology and Evolution* **12**, 604–615.

Reich, D., Green, R.E., Kircher, M., et al. (2010) Genetic history of an archaic hominin group from Denisova Cave in Siberia. *Nature* **468**, 1053–1060.

Reich, D., Patterson, N., Kircher, M., et al. (2011) Denosova admixture and the first modern human dispersals into Southeast Asia and Oceania. *American Journal of Human Genetics* **89**, 1–13.

Relethford, J.H. (1997) Hemispheric difference in human skin color. *American Journal of Physical Anthropology* **104**, 449–457.

Relethford, J.H. (2000) Human skin color diversity is highest in sub-Saharan African populations. *Human Biology* **72**, 773–780.

Relethford, J.H. (2002) Apportionment of global human genetic diversity based on craniometrics and skin color. *American Journal of Physical Anthropology* **118**, 393–398.

Relethford, J.H. (2004) Global patterns of isolation by distance based on genetic and morphological data. *Human Biology* **76**, 499–513.

Relethford, J.H. (2007) Population genetics and paleoanthropology. In Henke, W. and Tattersall, I. (eds) *Handbook of Paleoanthropology. Volume I: Principles, Methods and Approaches*. Springer-Verlag, Berlin, pp. 621–641.

Relethford, J.H. (2008) Genetic evidence and the modern human origins debate. *Heredity* **100**, 555–563.

Relethford, J.H. (2009) Race and global patterns of phenotypic variation. *American Journal of Physical Anthropology* **139**, 16–22.

Relethford, J.H. (2010) Population-specific deviations of global human craniometric variation from a neutral model. *American Journal of Physical Anthropology* **142**, 105–111.

Relethford, J.H. (2012) *Human Population Genetics*. Wiley-Blackwell, Hoboken, NJ.

Relethford, J.H. (2013a) Genetic drift and the population history of the Irish Travellers. *American Journal of Physical Anthropology* **150**, 184–189.

Relethford, J.H. (2013b) *The Human Species: An Introduction to Biological Anthropology*, 9th edn. McGraw-Hill, New York.

Reno, P.L., McCollum, M.A., Meindl, R.S., and Lovejoy, C.O. (2010) An enlarged postcranial sample confirms *Australopithecus afarensis* dimorphism was similar to modern humans. *Philosophical Transactions of the Royal Society B* **365**, 3355–3363.

Rightmire, G.P. (1990) *The Evolution of* Homo erectus. Cambridge University Press, Cambridge.

Robins, A.H. (1991) *Biological Perspectives on Human Pigmentation*. Cambridge University Press, Cambridge.

Rodman, P.S. and McHenry, H.M. (1980) Bioenergetics and the origin of human bipedalism. *American Journal of Physical Anthropology* **52**, 103–106.

Roebroeks, W. and Villa, P. (2011) On the earliest evidence for habitual use of fire in Europe. *Proceedings of the National Academy of Sciences* **108**, 5209–5214.

Rohde, D.L., Olsen, S., and Chang, J.T. (2004) Modelling the recent common ancestry of all living humans. *Nature* **431**, 562–566.

Roychoudhury, A.K. and Nei, M. (1988) *Human Polymorphic Genes: World Distribution*. Oxford University Press, New York.

Ruff, C.B. and Hayes, W.C. (1983) Cross-sectional geometry of Pecos Pueblo femora and tibiae—A biomechanical investigation: I. Method and general patterns of variation. *American Journal of Physical Anthropology* **60**, 359–381.

Ruff, C.B., Trinkaus, E., and Holliday, T.W. (1997) Body mass and encephalization in Pleistocene *Homo*, *Nature* **387**, 173–176.

Russell, M.D. (1985) The supraorbital torus: "A most remarkable peculiarity." *Current Anthropology* **26**, 337–360.

Sarich, V.M and Wilson, A.C. (1966) Quantitative immunochemistry and the evolution of primate albumins: Micro-complement fixation. *Science* **154**, 1563–1566.

Sarich, V.M. and Wilson, A.C. (1967a) Immunological time scale for hominid evolution. *Science* **158**, 1200–1203.

Sarich, V.M. and Wilson, A.C. (1967b) Rates of albumin evolution in primates. *Proceedings of the National Academy of Sciences* **58**, 142–148.

Savage-Rumbaugh, S. and Lewin, R. (1994) *Kanzi: The Ape at the Brink of the Human Mind*. John Wiley & Sons, New York.

Scally, A., Dutheil, J.Y., Hillier, L.W., *et al.* (2012) Insights into hominid evolution from the gorilla genome sequence. *Nature* **483**, 169–175.

Scarre, C. (2013) The world transformed: From foragers and farmers to states and empires. In Scarre, C. (ed.) *The Human Past: World Prehistory & the Development of Human Societies*, 3rd edn. Thames & Hudson, London, pp. 176–199.

Schick, K.D. and Toth, N. (1993) *Making Silent Stones Speak: Human Evolution and the Dawn of Technology*. Simon & Schuster, New York.

Schoenemann, P.T. (2006) Evolution of the size and functional areas of the human brain. *Annual Review of Anthropology* **35**, 379–406.

Schoenemann, P.T. (2013) Hominid brain evolution. In Begun, D.R. (ed.) *A Companion to Paleoanthropology*. Wiley-Blackwell, Chichester, UK, pp. 136–164.

Schoenemann, P.T., Budinger, T.F., Sarich, V.M., and Wang, W.S.-Y. (2000) Brain size does not predict general cognitive ability within families. *Proceedings of the National Academy of Sciences* **97**, 4932–4937.

Schow, D.J. and Frentzen, J. (1986) *The Outer Limits: The Official Companion*. Ace, New York.

Schrenk, F. (2013) Earliest *Homo*. In Begun, D.R. (ed.) *A Companion to Paleoanthropology*. Wiley-Blackwell, Chichester, UK, pp. 480–496.

Shen, G., Gao, X., Gao, B., and Granger, D.E. (2009) Age of Zhoukoudian *Homo erectus* determined with $^{26}Al/^{10}Be$ burial dating. *Nature* **458**, 198–200.

Shimelmitz, R., Kuhn, S.L., Jelinek, A.J., *et al.* (2014) "Fire at will": The emergence of habitual fire use 350,000 years ago. *Journal of Human Evolution* **77**, 196–203.

Smith, T.M., Tafforeau, P., Reid, D.J., *et al.* (2010) Dental evidence for ontogenetic differences between modern humans and Neanderthals. *Proceedings of the National Academy of Sciences* **107**, 20923–20928.

Soodyall, H. (2013) Lemba origins revisited: Tracing the ancestry of Y chromosomes in South African and Zimbabwean Lemba. *South African Medical Journal* **103**, 1009–1013.

Spoor, F., Leakey, M.G., Gathogo, P.N., et al. (2007) Implications of the new early *Homo* fossils from Ileret, east of Lake Turkana, Kenya. *Nature* **448**, 688–691.

Stanford, C. (2003) *Upright: The Evolutionary Key to Becoming Human.* Houghton-Mifflin, Boston.

Story, R. (1976) *The Space-Gods Revealed: A Close Look at the Theories of Erich von Däniken.* Harper & Row, New York.

Straus, W.L., Jr and Cave, A.J.E. (1957) Pathology and the posture of Neanderthal Man. *The Quarterly Review of Biology* **32**, 348–363.

Stringer, C. and McKie, R. (1996) *African Exodus: The Origins of Modern Humanity.* Henry Holt and Co., New York.

Stringer, C.B., Trinkaus, E., Roberts, M.B., et al. (1998) The Middle Pleistocene human tibia from Boxgrove. *Journal of Human Evolution* **34**, 509–547.

Sutikna, T., Tocheri, M.W., Morwood, M.J., et al. (2016) Revised stratigraphy and chronology for *Homo floresiensis* at Liang Bua in Indonesia. *Nature* **532**, 366–369.

Suwa, G., Kono, R.T., Simpson, S.W., et al. (2009) Paleobiological implications of the *Ardipithecus ramidus* dentition. *Science* **326**, 94–99.

Sykes, B.C., Mullis, R.A., Hagenmuller, C., et al. (2014) Genetic analysis of hair samples attributed to yeti, Bigfoot and other anomalous primates. *Proceedings of the Royal Society B* **281**, 20140161. http://dx.doi.org/10.1098/rspb.2014.0161.

Tattersall, I. (2009) *The Fossil Trail: How We Know What We Think We Know about Human Evolution*, 2nd edn. Oxford University Press, New York.

Thieme, H. (2000) Lower Palaeolithic hunting weapons from Schöningen, Germany—The oldest spears in the world. *Acta Anthropologica Sinica* **19**, suppl., 140–147. (Reprinted in Ciochon, R.L. and Fleagle, J.G. (eds) (2006) *The Human Evolution Source Book*, 2nd edn. Pearson Prentice Hall, Upper Saddle River, NJ, pp. 440–445.)

Thomas, M.G., Parfitt, T., Weiss, D.A., et al. (2000) Y chromosomes travelling south: The Cohen Modal Haplotype and the origin of the Lemba—the "Black Jews of Southern Africa." *American Journal of Human Genetics* **66**, 674–686.

Tishkoff, S.A., Reed, F.A., Friedlaender, F.R., et al. (2009) The genetic structure and history of Africans and African Americans. *Science* **324**, 1035–1044.

Tishkoff, S.A., Reed, F.A., Ranciaro, A., et al. (2007) Convergent adaptation of human lactase persistence in Africa and Europe. *Nature Genetics* **39**, 31–40.

Trinkaus, E. (1985) Pathology and the posture of the La Chapelle-aux-Saints Neandertal. *American Journal of Physical Anthropology* **67**, 19–41.

Trinkaus, E., Churchill, S.E., Ruff, C.D., and Vandermeersch, B. (1999) Long bone shaft robusticity and body proportions of the Saint-Césaire 1 Châtelperronian Neanderthal. *Journal of Archaeological Science* **26**, 753–773.

Trinkaus, E. and Shipman, P. (1992) *The Neandertals: Changing the Image of Mankind.* Alfred A. Knopf, New York.

Ungar, P.S., Grine, F.E., and Teaford, M.F. (2008) Dental microwear and diet of the Plio-Pleistocene hominin *Paranthropus boisei*. *PLoS One* **3**(4): e2044. doi:10.1371/journal.pone.0002044.

Villmoare, B., Kimbel, W.H., Seyoum, C., *et al.* (2015) Early *Homo* at 2.8 Ma from Ledi-Geraru, Afar, Ethiopia. *Science* 347, 1352–1355.

Von Däniken, E. (1970) *Chariots of the Gods.* Berkeley Books, New York.

Wainscoat, J. (1987) Human evolution: Out of the Garden of Eden. *Nature* 325, 13.

Walker, A. and Shipman, P. (2005) *The Ape in the Tree: An Intellectual and Natural History of Proconsul.* Harvard University Press, Cambridge, MA.

Walker, A. and Teaford, M. (1989) The hunt for *Proconsul. Scientific American* 260(1), 76–82.

Wall, J.T., Yang, M.A., Jay, F., *et al.* (2013) Higher levels of Neanderthal ancestry in East Asians than in Europeans. *Genetics* 194, 199–209.

Weaver, T.D., Roseman, C.C., and Stringer, C.B. (2007) Were Neandertal and modern human cranial differences produced by natural selection or genetic drift? *Journal of Human Evolution* 53, 135–145.

Weiner, S., Xu, Q., Goldberg, P., *et al.* (1998) Evidence for the use of fire at Zhoukoudian, China. *Science* 281, 251–253.

Weiss, K.M. (1984) On the number of members of the genus *Homo* who have ever lived, and some evolutionary implications. *Human Biology* 56, 637–649.

Westaway, M.C., Durband, A., and Collard, M. (2015) Down syndrome theory on Hobbit species doesn't hold to scrutiny. *The Conversation*, February 9. Available online at https://theconversation.com/down-syndrome-theory-on-hobbit-species-doesnt-hold-to-scrutiny-33375# (accessed July 13, 2016).

Westaway, M.C., Durband, A.C., Groves, C.P., and Collard, M. (2015) Mandibular evidence supports *Homo floresiensis* as a distinct species. *Proceedings of the National Academy of Sciences* 112, E604–E605.

Wheeler, P.E. (1991) The influence of bipedalism on the energy and water budgets of early hominids. *Journal of Human Evolution* 21, 117–136.

White, T.D., Ambrose, S.H., Suwa, G., *et al.* (2009) Macrovertebrate paleontology and the Pliocene habitat of *Australopithecus ramidus. Science* 326, 87–93.

White, T.D., Asfaw, B., Beyene, Y., *et al.* (2009) *Ardipithecus ramidus* and the paleobiology of early hominids. *Science* 326, 75–86.

White, T.D., Asfaw, B., DeGusta, D., *et al.* (2003) Pleistocene *Homo sapiens* from Middle Awash, Ethiopia. *Nature* 423, 742–747.

White, T.D., WoldeGabriel, G., Asfaw, B., *et al.* (2006) Asa Issie, Aramis and the origin of *Australopithecus. Nature* 440, 883–889.

Whiten, A., Goodall, J., McGrew, W.C., *et al.* (1999) Culture in chimpanzees. *Nature* 399, 682–685.

Wolpoff, M.H. (1999) *Paleoanthropology.* McGraw-Hill, Boston.

Wolpoff, M.H., Thorne, A.G., Smith, F.H., *et al.* (1994) Multiregional evolution: A world-wide source for modern human populations. In Nitecki, M.H. and Nitecki, D.V. (eds) *Origins of Anatomically Modern Humans.* Plenum Press, New York, pp. 175–199.

Wolpoff, M.H., Xinzhi, W., and Thorne, A.G. (1984) Modern *Homo sapiens* origins: A general theory of hominid evolution involving the fossil evidence from East Asia. In Smith, F.H. and Spencer, F. (eds) *The Origins of Modern Humans*. Liss, New York, pp. 411–483.

Wood, B. (2012) Facing up to complexity. *Nature* **488**, 162–163.

Wood, B. (2014) Fifty years after *Homo habilis*. *Nature* **508**, 31–32.

Zolliker, C.P.E., Ponce de León, M.S., Lieberman, D.E., *et al.* (2005) Virtual cranial reconstruction of *Sahelanthropus tcadensis*. *Nature* **434**, 755–759.

INDEX

ABO blood group system, 17, 249
Acheulian tools, 129–132, 137, 243
Acosta, Friar Joseph de, 192
adaptation, 13, 16–22, 26
  cultural, 252
  to savanna environment, 77–79
adenine, 163
Afar, 81
African Americans, 227–231, 234, 236, 237
African apes, 45, 46, 50–56, 61, 62, 72–76
  see also apes
African replacement model, 162, 170
Africans, 195, 224, 225, 228
aggression, 86, 90
Agricultural Revolution, 203
agriculture, 2, 203–206, 238, 244, 245, 252
AIDS, *see* HIV/AIDS
Aiello, Leslie, 119
albumin protein analysis, 54, 55
alleles, 18, 19, 22, 27, 196, 228, 238, 246, 251
  dominant, 246, 247, 249
  frequencies, 20–21, 229, 249

hemoglobin *S*, 235, 245
lactase persistence, 244–245
lactase restriction, 244
recessive, 246–249
sickle cell, 235, 236, 239, 245
American Sign Language (ASL), 124–125
anagenesis, 40–44, 91, 98, 109, 111, 113
Andaman Islanders, 145–146
Andrews, Roy Chapman, 159
anemia, 234–239
anthropology 121
antibiotics, 211–217, 246, 252
apes,
  African, 45, 46, 50–56, 61, 62, 72–76
  Asian, 50, 55
  brain size, 115
  cranial base, 155
  differences from humans, 46–48
  diversity in, 58
  and human evolution, 39–44, 101–106
  intelligence, 66
  jaws, 47–48

50 Great Myths of Human Evolution: Understanding Misconceptions about Our Origins,
First Edition. John H. Relethford.
© 2017 John Wiley & Sons, Inc. Published 2017 by John Wiley & Sons, Inc.

apes (cont'd)
  not among first mammals, 34
  as primates, 31
  split from humans, 51–56, 178
  teeth, 47–48, 50, 65
  use of language 124–125
Aramis site, 78
archaeology, 250
  experimental, 210
Arctic Northeast Asians, 195
Ardey, Robert, 86, 88
Ardipithecus,
  bipedalism, 75, 78–80, 86
  brain size, 86, 105
  early hominins, 56, 96
  fossils, 73–75, 86, 103
  teeth, 65, 78, 103, 105
Ardipithecus kadabba, 64, 93, 98, 103
  bipedalism, 76
Ardipithecus ramidus, 64, 65, 74, 76, 79, 93, 98, 103
  oldest definite hominin, 56
argon dating, 36
Aristotle, 30
artificial selection, 14, 16
Asian apes, 50, 55
Asians, 224
assimilation model, 162, 170
asteroid impact, 250
atlatl, 143, 152
Australasians, 195
Australia, 128, 144, 158, 161, 167, 174, 191, 198, 199
Australian Aborigines, 174, 224, 226
Australopithecus (genus),
  bipedalism, 86, 103, 106
  brain size, 86, 90, 110–111, 116, 120, 133, 178–179, 240
  cranial capacity, 179
  diet, 95, 96
  as early human ancestor, 70, 105–106, 109–111
  evidence against Piltdown Man as human ancestor, 71
  evolution in Africa, 160
  face size, 110
  and fire use, 135–136
  hyoid bone, 156
  as a "killer ape", 2, 85–90
  and language, 124
  meaning of name, 65
  species, 94, 105
  teeth, 65, 103, 104, 110
  tool use, 85, 89, 128
Australopithecus afarensis, 76, 79–85, 93, 98, 103–106, 114
Australopithecus africanus, 68, 81, 86, 91, 93, 94, 97, 104, 105
Australopithecus anamensis, 65, 76, 81, 93, 98, 103
Australopithecus bahrelghazali, 98
Australopithecus deyiremeda, 93, 98, 101
Australopithecus garhi, 89, 93, 94, 98, 105
"*Australopithecus prometheus*", 86
Australopithecus sediba, 93, 94, 98, 105
Aztecs, 143

baboons, 43
  skulls, 87
bacteria, 8, 214, 216
balancing selection, 238
Basque population, 17
behavior,
  group specific, 123
  influences on, 118
  learned, 120, 122, 123
  symbolic, 152, 154, 240
Beringia, 193
Bering Land Bridge, 59, 193
Bering Strait, 193
beta chain, 235
bicuspids, 48
Bigfoot, 56–60
big-game hunting, 139, 141, 143, 152, 193
"bigger is always better" myth, 22–26

biochemical analysis, 53, 54
biocultural evolution, 244
biological age, 84
biomechanical model of brow-ridge
  formation, 147, 148
biostratigraphy, 36
bipedalism,
  antiquity of, 65
  of "Ardi", 74, 75
  in *Ardipithecus*, 75, 78–80, 86
  in *Australopithecus afarensis*, 82
  benefits and costs, 27
  and brain size, 63, 66, 76
  in chimpanzees, 79
  disadvantages, 79–80
  emergence in apes, 34
  energy efficiency, 77, 79
  evolution, 63–65, 75–80
  facultative, 74, 75, 80
  and fossil evidence, 66, 77
  in *Gigantopithecus*, 56, 59, 60
  in hominins, 56, 60, 75–80,
    105–106
  inference from jaws, 60
  and knuckle walking, 73
  in Neandertals, 149, 151, 170
  obligate, 62, 75, 80
  origin, 76
  in *Orrorin*, 76
  pelvic adaptations, 59, 62, 65
  predating human traits, 76
  in *Ramapithecus* 44
  specialization, 62
  and tool use, 49, 76, 77, 80
  as unique human trait, 62
Bismarck Archipelago, 200
Black Death (bubonic plague),
  8–9, 205
"Black Jews", 233
"Black Skull", 104, 105
blond hair, 246–249
blood groups, 17, 194
blood markers, 226
blood proteins, 53

boat travel, 199
Boaz, Noel, 141
body size
  and anagenesis, 40
  in *Australopithecus afarensis*, 80–85
  and brain size, 62, 116
  and environmental circumstances,
    22, 24
  evolution, 22–26
  relationship with organ size, 119
Bolnick, Deborah, 196
bonobos, 45, 46, 51, 61, 62, 72,
  73, 75
Boule, Marcellin, 150, 151, 152
Bouri site, 89, 160
bows and arrows, 143, 152
Boyd, William, 224
brachydactyly, 248
brain,
  complexity, 118
  development, 154, 177
  size, *see* brain size
  structure, 117
Brain, C. E., 88
"brain first" model of human
  evolution, 67, 70, 71
brain size, 178–183
  absolute, 116
  in apes, 115
  of "Ardi", 74
  in *Ardipithecus*, 86, 105
  in *Australopithecus*, 86, 90,
    110–111, 116, 120, 133,
    178–179, 240
  and bipedalism, 63, 66, 76
  of "Black Skull", 105
  and body size, 62, 116
  and cognitive ability/intelligence,
    24, 49, 239–242
  costs, 118–120
  effect on childbirth, 80, 119–120
  estimation, 65
  and evolution, 61, 66–71, 104, 240
  and fitness, 24, 25

brain size (*cont'd*)
  and the fossil record, 23, 239
  future, 251
  and gut size, 119
  in hominins, 104, 105, 110
  in *Homo* (genus), 5, 23, 65, 115–120, 125–126, 178
  in *Homo erectus*, 91, 110, 112, 120
  in *Homo habilis*, 110, 120, 133
  in *Homo heidelbergensis*, 126
  in *Homo rudolfensis*, 112
  in humans, 9, 11, 23–25, 60–62, 66–71, 128, 158, 177, 178, 240
  and island dwarfism, 182
  and latitude, 242
  and natural selection, 17
  in Neandertals, 117, 127, 149, 152, 170, 177, 178
  see also brain
Broom, Robert, 69
brow ridges, 84, 112, 143–148, 158
Bryc, Katarzyna, 230
bubonic plague, *see* Black Death
"bush" metaphor, 91, 110
butchering, 132, 134, 139, 140, 141

cancer, 204, 214, 222
canine teeth, 46–50
  in apes, 65
  in *Ardipithecus*, 65
  in *Australopithecus* (genus), 65
  in *Australopithecus afarensis*, 82
  in human evolution, 103, 104
  replacement by tools, 48
  size, 54, 62–63, 65, 82, 105–106
  variance among apes, 50
Cann, Rebecca, 167
carbon, 136, 138
carbon-14 dating, 35, 152
Cartmill, Matt, 57
Cavalli-Sforza, Luca, 195
Cave, A. J. E., 151
caves, 89, 134
cavities, 204

Cenozoic Era, 34
charcoal, 136
Chardin, Teilhard de, 71
cheek bones, 94
chemical bases, 163
chemical staining, 136
chewing, 94
childbirth, 80, 119–120, 251
chimpanzees,
  aggression, 90
  bipedalism, 79
  climbing, 62
  knuckle walking, 73, 75, 77
  use of language, 124
  "rain dance", 122
  relationship to humans, 40, 44, 45, 46, 51, 55, 61, 62, 72–76
  sexual dimorphism, 84
  tool use, 63, 120–123
China, 58
chopping tools, 130
chromosomes, 9, 163, 190, 197
chronological age, 83, 84
Ciochon, Russell, 141
cities, 210
cladogenesis, 41–44, 98, 109
climate change, 77, 203, 251
climbing,
  in *Ardipithecus*, 79
  in chimpanzees and bonobos, 62
  suspensory, 75
Clovis
  burial site, 197
  culture, 193
  DNA analysis from, 197
  stone tools, 193, 194
Clovis point, 193
coalescence analysis, 165, 168
cockroaches, 30
cognitive ability, 126, 239–242
  and brain size, 24, 49, 239–242
  see also mental ability
common ancestors, 163–169, 187, 188, 189

common cold, 204
comparative molecular analysis, 54
competition, 23, 27
computer analysis, 175
computer simulations, 20–21
computer technology, 100
consensus, 2, 4
contamination, 213
Cook Island, 201
Coon, Carleton, 224
correlation analysis, 240
correlations, 241
cranial base, 155
cranial capacity
   of *Australopithecus* (genus), 179
   of *Australopithecus afarensis*, 104
   of early hominins, 65
   as function of geologic age, 116–117
   of *Homo* (genus), 179
   of *Homo erectus*, 104, 117
   of *Homo rudolfensis*, 114
   of *Kenyanthropus platyops*, 114
   of "LB1", 179
   as measure of volume of skull interior, 116
cranial expansion, 65
cranial measures, 223, 226
cranial reconstructions, 157
cranial shape, 177, 181
Crawford, Michael, 233
Crelin, Edmund, 155
Cretaceous period, 34
cretinism, 181
Croatia, 171
Cro-Magnon rock shelter, 159
CT scanning, 65, 100
cultural adaptation, 244, 252
cultural change, 2, 243–246, 252
cultural evolution, 243, 244
cultural identity, 231–234
culture, 2, 120, 131, 158, 231–234, 243, 244, 252
   Clovis, 193

Lapita, 200
Solutrean, 194
transmission of, 118
cusps, 61, 103–104
cut marks, 140–142
cytosine, 163

dairy farming, 244, 245
Dart, Raymond, 68, 70, 81, 86–89, 99, 136
Darwin, Charles, 10–14, 48, 76, 102, 144–145, 158, 224
dating, 35–37, 70, 152
Dawson, Charles, 67, 71
death rates, 212, 213, 215
deer remains, 139
deforestation, 246
demographic changes, 245
Denisova Cave, 174
Denisovans, 174, 176
dental health, 204
dental hypoplasia, 205
dental remains, 46, 49, 52
   see also jaws; teeth
derived traits, 30–31, 53
diabetes, 204
Diamond, Jared, 203
diarrheal diseases, 213, 216
diastema, 103
Diego blood group system, 17
diet,
   of *Ardipithecus*, 78
   of *Australopithecus*, 95, 96
   changes in, 119, 204, 245
   human, 244
   meat in, 139–143
   of *Paranthropus*, 95–96
dietary diversity, 204
dinosaurs, 33–37, 180, 250
   eggs, 159
directional selection, 31
diseases, 181, 206, 251, 252
   endemic, 205
   epidemic, 205

diseases (*cont'd*)
   infectious, 8, 203–205, 211, 213–216, 244–245
   noninfectious, 204, 213–214
diversity, 28, 44, 58, 90
DK site, 133
Dmanisi site, 99, 137, 181
DNA
   analysis, 100, 176, 201, 202
   difficulty of preserving, 176
   markers, 198, 200–202, 223, 226, 228, 229
   microsatellite, 201, 202
   mitochondrial, 163–168, 171, 174, 190, 195, 197, 201, 233
   and mutation, 14
   Neandertal, 157, 171, 172, 173
   nuclear, 171, 172, 174
   and the relationship of humans to Neandertals, 161–162
domestication, 13, 16
dominant alleles, 246, 247, 248, 249
Down syndrome, 182
Druze, 196
Dubois, Eugene, 159
Duffy blood group system, 17, 239
dwarf elephants, 26, 180

Early Pleistocene, 67, 70
East Africa, 36, 99, 100, 105, 113
   earliest stone tools in, 65
   early human evolution in, 36
East Asia, 58
East Asians, 224
Easter Island, 198, 199, 202, 208
Ebola, 216
Egyptian pyramids, 206–210
electron spin resonance dating, 36
El Sidrón site, 156, 157
endocasts, 116, 154
energy efficiency, 77, 79
enteritis, 213
environment
   adaptation to, 26

and body size, 22, 24
and natural selection, 12, 13, 28
environmental change, 245, 253
Eoanthropus dawsoni, 68
epidemiologic transition, 214, 215
Ethiopia, 64, 65, 78, 81, 89, 104
eumelanin, 248
European Americans, 234
Europeans, 195, 224
evolution,
   and adaptation, 16–22
   biocultural, 244
   of bipedalism, 63–65, 75–80
   and body size, 22–26
   of the brain, 61, 66–71, 104, 117–118, 240
   "brain first" model, 67, 70, 71
   "bush" metaphor, 91, 100
   cultural, 243, 244
   cumulative nature of, 5
   extraterrestrial influence, 2, 206–211
   and face size, 61
   of hominins, 45–46
   of *Homo erectus*, 91, 93, 97, 98, 103, 109–111, 113, 115, 126
   of humans, 39–44, 60–66, 101–106, 187, 243–246
   importance of fossils, 175–178
   "ladder" metaphor, 90–96, 110
   misconceptions about, 2
   and mutation, 17
   and natural selection, 17, 18, 42, 249
   nonadaptive, 18, 19
   prediction of, 249–253
   random, 12–16, 251–252
   "scars of", 80
   and selective breeding, 13
   of skin color, 220
   of stone tool technology, 125–132
   as theory, 7–11
evolutionary forces, 17–18, 42, 249
excavations, 210
expensive-tissue hypothesis, 119

experimental archaeology, 210
"express train" model of Polynesian dispersal, 200–202
extinction, 10–11, 26, 28, 33, 43–44, 51, 56, 111, 162, 196, 247
  of ape species, 58
  of Denisovans, 174
  of dinosaurs, 250
  of *Gigantopithecus*, 58–59
  due to global annihilation, 253
  of *Homo floresiensis*, 183
  of *Homo rudolfensis*, 113
  K/T (Cretaceous/Tertiary), 28
  of mammal-like reptiles, 102
  of Neandertals, 97, 153, 162, 169–170, 173
  of *Paranthropus*, 90, 96
extraterrestrials, 2, 206–211
eye color, 19

face,
  of "Black Skull", 105
  of early hominins, 111
  evolution of, 11, 61, 147
  of *Homo* (genus), 5, 115, 126
  of *Homo erectus*, 112
  of *Homo habilis*, 110, 112
  of *Homo heidelbergensis*, 126–127
facial anatomy, 114
facts, 7, 8, 9
facultative bipedalism, 74, 75, 80
farming, 244
Feder, Kenneth, 71, 208
Fertile Crescent, 203
fire, 86, 87, 135–138
first Americans, 192–198
fission-track dating, 36
fitness, 12, 24, 25, 31
flakes, 120, 128
flint, 137, 138, 142
Flores, 179, 180
fluorine dating, 70
fluting, 193
folate levels, 222

food resources, 24, 27, 251
food sharing, 77, 134
food surpluses, 210
foramen magnum, 64, 68, 74
forkhead box protein P2, 157
fossils,
  anagenesis in 43–44
  Ardipithecus, 73–75, 86, 103
  behavioral correlates from 118
  and bipedalism, 66, 77
  and brain size, 23, 239
  cladogenesis in, 43–44
  evidence for speech, 157
  extent of the record, 99–100
  and extinction, 28
  and hominin evolution, 56, 86
  Homo heidelbergensis, 162
  and human evolution, 61, 175–178
  imperfect nature of record, 102
  Neandertal, 149, 157, 171
  and origins of *Homo*, 5, 34
  preservation of unique human traits in, 63
  and relationship of apes to humans, 51, 52, 55
  and relationship of *Ramapithecus* to humans, 49–50, 52
  of tetrapods, 31
  and tool technology, 126
  of transitional forms, 101–106
FOXP2 gene, 157
Friedlaender, Jonathan, 201

Garn, Stanley, 224
gathering, 204–206, 252
Gehser Benot Ya'aqov (GBY), 137
genealogical ancestry, 190
gene flow, 173, 246
  effects of reduction, 43
  as an evolutionary force, 18, 42, 249
  introduction of new alleles, 27
genetic analysis, 53, 54, 163, 175, 239
genetic ancestry, 190, 231–234
genetic correlations, 241

Index | **279**

genetic distance, 195
genetic drift, 149, 168, 171, 196, 201, 246, 250, 252
   and allele frequency, 21
   and anagenesis, 40
   as an evolutionary force, 18, 42, 249
   definition, 18
   example, 20
   in natural theory of molecular evolution, 22
   nonadaptive evolution, 19
   as random fluctuations, 16
genetic engineering, 250
genetic evidence, 49–50, 200
genetic markers, 194, 195, 201, 223
genetic technologies, 171
genetic variation, 12–14
genoptype frequencies, 249
geographic evidence, 192
geographic races, 225
geologic age, 117
geologic dating, 99
germ theory of infectious disease, 215
gibbons, 55, 84
Gigantopithecus, 56–60
Giza, 208, 209
global annihilation, 253
global warming, 246
gnawing, 140
Goodall, Jane, 121–122
Goodman, Morris, 54
gorillas, 45, 46, 51, 55, 61, 72, 73, 75, 84
Gould, Stephen Jay, 91
Gran Dolina, 141
gravity, 10
Great Rift Valley, 100, 133
Green, Richard, 172, 173
grooming, 123
Grotte du Renne site, 152
group-specific behavior, 123
guanine, 163
Gullah, 229
gut size, 119, 120

Hadar site, 81
Haeckel, Ernest, 159, 161
hair color, 246–249
hairiness, 54, 61, 63
Halakha, 234
half life, 35, 36
hand axes, 130
haplogroups, 195, 196
haplotypes, 164, 195
Harappa site, 210
Hardy, Godfrey, 248, 249
Hardy–Weinberg law, 249
Hawaii, 198, 199
health, 2, 203–206, 243
heart disease, 204, 213, 214
hearths, 135
heat, 222
height, 226
hemoglobin, 235, 238, 245
heterozygotes, 238
Heyerdahl, Thor, 198, 199, 200, 202
Hierakonpolis site, 210
HIV/AIDS, 216
"Hobbit", 99, 178, 179, 183
Holloway, Ralph, 154
home bases, 134
hominids, *see* hominins
hominins,
   anatomy, 99
   bipedalism, 56, 60, 75–80, 105–106
   brain size, 104, 105, 110
   coexistence of different species, 90
   definition, 46, 56
   discovery, 97–101
   evolution, 45–46
   first, 105
   genera, 103
   proposed species, 93
   split from apes, 178
Homo (genus), 105, 106
   brain size, 5, 23, 65, 115–120, 125–126, 178
   cranial capacity, 179
   descendants of *Australopithecus*, 96

face size, 5, 115, 126
first fossils, 34
language ability, 117
origins, 5
species, 2, 109–115
stone tool technology, 5, 125–132
teeth, 104, 115, 126
Homo antecessor, 142
Homo erectus,
  age, 111
  brain size, 91, 110, 112, 120
  brow ridges, 144
  cranial capacity, 104, 117
  discovery of first specimen, 159
  dispersion of 167
  division into further species, 92
  at the Dmanisi site, 99
  evidence in Africa, 160
  evolution, 91, 93, 97, 98, 103, 109–111, 113, 115, 126
  face, 112, 127
  fire use, 135–138
  gut size, 119, 120
  hunting, 139–143
  relationship to *Homo floresiensis*, 180
  relationship to *Homo heidelbergensis*, 32
  relationship to *Gigantopithecus*, 58
  as scavenger, 141
  use of caves for shelter, 134
  use of tools 129–131
Homo ergaster, 98, 111
Homo floresiensis, 93, 98, 179–183
Homo habilis, 93, 97, 104, 109–113, 117, 120, 126, 129, 131, 179, 181
  brow ridges, 146
  creation of shelter, 132–135
  evidence in Africa, 160
Homo heidelbergensis, 93, 98, 104, 117, 126, 127, 131, 137–138, 142, 144, 146, 149, 154–158, 162
  as big-game hunter, 141
  relationship to *Homo erectus*, 32

relationship to *Homo sapiens*, 32
homoplasy, 73
Homo naledi, 98, 101, 115
Homo neanderthalensis, 93, 149
  see also Neandertals
Homo rudolfensis, 93, 104, 111–114, 117, 126, 129, 131
Homo sapiens, 91, 93, 104, 109, 111, 117, 126, 149, 158
  "archaic", *see Homo heidelbergensis*
  burial of dead, 132
  evidence in Africa, 160, 161
  relationship to *Homo heidelbergensis*, 32
  relationship to Neandertals, 127–128, 162
  skulls, 160
  tool use, 131
  use of projective weapons, 143
  see also Homo sapiens sapiens; humans
Homo sapiens neanderthalensis, 149
  see also Neandertals
Homo sapiens sapiens, 109, 224
honing complex, 48
Hooton, Earnest, 150
Howell, F. Clark, 159
humans,
  appearance in Eurasia, 158–162
  appearance in sub-Saharan Africa, 158
  "brain first" model of evolution, 67, 70, 71
  brain size, 9, 11, 23–25, 60–62, 66–71, 128, 158, 177, 178, 240
  coexistence with Neandertals, 161, 169–175
  continuing evolution, 243–246
  differences from apes, 46–48
  dispersal from Africa, 128
  and dinosaurs, 33–37
  evolution from apes, 39–44, 101–106

humans (*cont'd*)
  evolution of bipedalism, 75–80
  evolution of traits, 60–66
  and extinction of *Homo floresiensis*, 183
  evolution from *Homo heidelbergensis*, 158
  face size, 11
  and first mammals, 34
  first modern, 35
  future evolution, 187, 249–253
  increase in body size, 244
  jaws, 47–48
  "ladder" theory of evolution, 90–96, 110
  New World primates, 53–54
  as obligate bipeds, 75
  and Old World primates, 53–55
  as primates, 31
  and *Ramapithecus*, 44–51
  relationship to chimpanzees, 40, 44, 45, 46, 51, 55, 61, 62, 72–76
  skulls, 158
  split from apes, 51–56
  teeth, 9, 47–48
  tool use, 152
  transitional nature of evolution, 101–106
  variation, 223–227
  waste, 205
  see also Homo sapiens; *Homo sapiens sapiens*
hunting, 89, 132, 134, 137, 139–143, 152, 154, 204–206, 252
hybridization, 43
hyenas, 88, 140
hygiene, 215
hyoid bones, 156
hypoglossal canal, 155, 156
hypotheses
  different from conclusions, 132
  about natural selection, 11
  and the scientific method, 3, 4, 9–10
  and theory, 8, 9

identical ancestors point, 191
immunological distance, 55
immunological reactions, 54
impact events, 28
inbreeding, 187, 188, 189
Incas, 199
incisors, 46–47, 62, 83, 149, 194
indels, 201, 202
India, 58
Indonesia, 199
Indonesians, 174
Industrial Revolution, 10, 246
infectious diseases, 8, 203–205, 211, 213, 215, 216, 244, 245
influenza, 211–214, 216
intelligence, *see* cognitive ability
intelligence test scores, 239–242
interbreeding, 43, 162–163, 167, 169–175
intermarriage, 228
interpretative bias, 133
iridium, 250
Irish Travellers, 232, 233
island dwarfism, 25, 26, 180, 181, 182
island gigantism, 26
isolation, 26, 43
isotope analysis, 78, 79, 95–96

Java, 180
jaws,
  of apes 47–48
  of *Australopithecus afarensis*, 114
  as clues to hominin ancestry, 46
  of *Gigantopithecus*, 58, 59, 60
  of humans, 47–48
  indicators of age, 83–84
  inference of bipedalism from, 60
  of *Paranthropus*, 94, 95
  of *Proconsul*, 52
  of *Ramapithecus*, 49–50
  U-shaped, 47, 49, 52
Jenner, Edward, 214
Jews, 233, 234

Johanson, Donald, 81
Jurassic period, 34

Kayser, Manfred, 202
Kebara Cave, 156
Kennewick Man, 197
Kenyanthropus platyops, 93, 98, 114
Khufu, 209
"killer ape", 2, 85–90
Kimura, Morton, 22
Klein, Richard, 134
knuckle walking, 45, 62, 72, 73, 75, 77
Kon-Tiki, 199
Krantz, Grover, 59, 60
Krogman, Wilton, 80
K/T (Cretaceous/Tertiary) extinction, 28

labor specialization, 210
labor surpluses, 210
La Chapelle-aux-Saints, 150, 151
lactase persistence, 244, 245
lactase restriction allele, 244
lactose intolerance, 244
Ladder of Being, *see Scala Naturae*
Lake Titicaca, 199
Lake Turkana, 95, 104, 111
language, 117–118, 120, 123–124, 155–157
  see also speech
Lapita culture, 200
Lapita pottery, 200
Laron's Syndrome, 181
larynx, 154, 155
latitude, 242
Law of Superposition, 34, 35
"LB1", 179, 180–182
Leakey, Louis, 51, 97, 99, 104, 133
Leakey, Mary, 99, 104, 133
Legionnaire's disease, 216
Lemba, 233, 234
Lemuria, 159
lemurs, 31
leopards, 89
Levallois tools, 130, 131, 243

Liang Bua cave, 179, 182
Libby, Willard, 35
Lieberman, Philip, 155
life expectancy, 206, 211–217
life-table analysis, 206
limb proportions, 110, 112
Linnaeus, Carl, 224
Livingstone, Frank, 238
Lomekwian tools, 128, 131
longevity, 177
lorises, 31
Lovejoy, Owen, 74, 75
Lower Paleolithic tools, 131
lower respiratory diseases, 214, 216
Lucy, 80–85
"lumpers", 92, 93, 98

Maasai, 173
magnetic pole, 37
magnetic resonance imaging (MRI), 240, 241
Makapansgat site, 86–89
malaria, 204, 216, 237, 238
mammal-like reptiles, 102
mammals, 34
mammoths, 193
mandibles, 87
manganese staining, 87
"man the hunter" model, 139
Marks, Jon, 231, 232
mastabas, 209
mastodons, 193
mastoid processes, 84
mathematical models, 175, 189
maturation, 83
measles, 204, 214, 215
meat, 139–143
medical advances, 211
medical imaging technology, 100, 116
Melanesia, 198, 200
Melanesians, 174, 200, 202
melanin, 219, 248
memory, 118

Index | 283

mental ability, 120
  see also cognitive ability
Mesozoic Era, 34
microcephaly, 181
Micronesia, 198, 202
microorganisms, 8, 205, 215
microsatellite DNA, 201, 202
Middle East, 203
Middle Easterners, 195
Middle Paleolithic tools, 131
Miocene epoch, 34, 58
missing links, 101
mitochondrial ancestors, 164, 165, 166
mitochondrial DNA (mtDNA), 163–168, 171, 174, 190, 195, 197, 201, 233
"Mitochondrial Eve", 163–169
MN blood group, 17, 18, 19
Moheno-Daro site, 210
molars, 46, 47, 60, 61, 84
molecular analysis, 245
molecular genetic analysis, 239
monkeys,
  as primates, 31
  hunted by raptors, 89
  leaf-eating, 78
  locomotion, 52, 73
  not among first mammals, 34
Morris, Desmond, 63
mosquitoes, 204, 217, 237–239
most recent common ancestor (MRCA), 164, 166, 190
most recent common genealogical ancestor, 190, 191
Mount Carmel, 159, 160
Mousterian culture, 152
MRI, see magnetic resonance imaging
multiregional evolution model, 160
mumps, 214
mutation, 195, 202, 236, 245–246, 250, 252
  as cause of variation, 14–15
  and DNA, 14

  as an evolutionary force, 17–18, 42, 249
  importance, 18
  introduction of new alleles, 27
  and natural selection, 15, 16
  random element in, 12, 15, 16
  and survival, 15
myths,
  and consensus, 2
  definition, 1–2

"naked apes", 63
Native Americans, 17, 192–198, 224
natural selection,
  action on variation, 14–15
  and anagenesis, 40
  and brain size, 17
  as compromise, 26
  continuing influences, 246
  and cultural adaptation, 244
  and the environment, 28
  as an evolutionary force, 17, 18, 42, 249
  and fitness, 24
  as hypothesis, 11
  importance 18
  and mutation, 15, 16
  prediction of, 251
  not random, 12, 16
  and sickle-cell anemia, 236
  and skin color, 17
  success of, 26–29
  and "survival of the fittest", 23
natural theory of molecular evolution, 22
Neandertals,
  as big game hunters, 152, 154
  bipedalism, 149, 151, 170
  brain size, 117, 127, 149, 152, 170, 177, 178
  brow ridges, 144–146
  burial of dead, 132, 150, 152, 153, 154
  Classic, 159

coexistence with *Homo sapiens*,
    127–128, 161, 169–175
  DNA from, 171, 172, 173
  evolution from *Homo
    heidelbergensis*, 127
  extinction, 169–170
  fossils, 149, 157, 171
  growth patterns, 177
  intelligence, 148–153
  language, 157
  locomotion, 2, 148–153
  mental abilities, 145
  Progressive, 159
  skin color, 176
  skulls, 127, 149
  speech, 153–157
  as subspecies of *Homo sapiens*, 162
  tool use, 152, 154, 161
  teeth, 149
  traits, 149
  trauma among, 143
  use of Levallois tools, 131
  as variant of *Homo sapiens*, 127
Neolithic, 203
New Guinea, 200
New Stone Age, 203
New Zealand, 198, 199
nomadic lifestyle, 204
noninfectious diseases, 204, 213–214
North Africa, 195, 203
Northeast Asians, 195
nuclear DNA, 171, 172, 174, 201
nutrition, 203
Nuttall, George, 53

obligate bipedalism, 62, 75, 80
occipital bun, 149
"Old Man" skeleton, 150, 151
Oldowan tools, 128–131, 133
Old Stone Age, 131
Olduvai Gorge, 110, 132, 133, 144
omnivorous diet, 78
Omo site, 160
orangutan, 50, 54, 70, 73

Orrorin, 56, 73, 74, 96
  possible bipedalism, 76
Orrorin tugenensis, 64, 93, 98
Osteodontokeratic Culture, 87, 88, 90
outrigger canoes, 199, 200
Owen, Richard, 145

Pacific Islanders, 195, 198, 224
Pakistan, 58
paleoanthropology, 5, 99, 177, 178
paleobotany, 99
Paleolithic, 131
paleontology, 250
Paleozoic Era, 28, 102
Paluxy River, 34
pandemics, 212
Papua New Guinea, 198
Paranthropus (genus), 90, 94, 96, 97,
    103, 105
Paranthropus aethiopicus, 92, 93,
    98, 104
Paranthropus boisei, 92, 93, 95,
    104, 105
Paranthropus robustus, 93, 105
Parra, Esteban, 229, 230
parsimony principle, 73, 75
Pasteur, Louis, 215
pedigree collapse, 189
peer review, 5–6
Peking Man, 137
pelvis, 59, 62, 65, 80, 84, 103, 120
penicillin, 215
peppered moth, 14, 15, 251
Permian period, 28
Peru, 199
pesticide resistance, 27
pharaohs, 209
phenotypes, 177
pheomelanin, 248
pigmentation, *see* skin color
Piltdown Man, 9, 66–71, 161
Plasmodum falciparum, 237
Pleistocene, 67, 70
pneumonia, 213, 216

polio, 215
pollen analysis, 99
pollution, 14, 251
polymerase chain reaction, 171
Polynesia, 198, 199, 208
Polynesians, 174, 198–202
population density, 205
population growth, 246
population pressure, 211
population size, 205, 246
post-cranial structure, 112
Potts, Richard, 134, 135
predators, 79, 251
prefrontal cortex, 117
premolars, 46, 47, 48, 103–104
primates, 30–31, 34, 43, 53–55
primitive traits, 30–31
probability theory, 20, 21
problem solving, 118, 120, 121
Proconsul, 51–53, 56, 58
Progressive Neandertals, 159
projective weapons, 143
public health, 211, 243
Punnett, Reginald, 248
pygmies, 179, 181

Qafzeh Cave, 160, 161
quadrupedalism, 60, 62

race, 223–227, 234, 237
radioactive decay, 35, 36
radioactive emissions, 35
rainforests, 77
Ramapithecus, 44–51, 62
raptors, 89
recessive alleles, 246–249
recessive traits, 246
red blood cell genetic markers, 223
reflectance spectrophotometers, 218
reproduction, 24, 29
Rhesus blood group system, 17
rhinoceroses, 139, 142
rice, 204
Rohde, Douglas, 189, 190, 191

Roma, 232, 233
Ruff, Christopher, 25
Russell, Mary Doria, 144

sagittal crest, 94, 105
Sahelanthropus, 56, 64, 73, 74, 76, 93, 96, 98
sampling error, 19
sanitation, 211, 213
Santanyana, George, 253
Sarich, Vincent, 54, 55
savanna, 76, 77
Scala Naturae, 30
scanning electron microscope (SEM), 139
scavenging, 88, 89, 119, 132, 139–143
Schaafhausen, Hermann, 144
Schoenemann, P. Thomas, 241, 242
Schöningen, 142
sciatic notch, 84
science,
 dynamic nature, 2–4, 44
 and hypotheses, 3, 4, 9–10, 71
 openness, 71
 as process of discovery, 101
science fiction, 206, 249
scientific method, 2, 3, 10
sedentary lifestyle, 204
selection, 27, 118, 246
 artificial, 15, 16
 directional, 31
 and fitness, 31
 stabilizing, 31
 see also natural selection
sex cells, 19
sex chromosomes, 197
sexual dimorphism, 84, 85
shelter, 132–135
Shkul Cave, 160, 161
shoulder blades, 87
shoulder joints, 61
Siberia, 193, 196, 242
sickle-cell alleles, 235, 236, 239, 245
sickle-cell anemia, 234–239

Sima de los Huesos site, 156
simulations, 20–21
Sivapithecus, 50
skeletal age, 83
skeletal analysis, 194
skin cancer, 222
skin color, 16, 17, 176, 217–222, 237, 247
skin reflectance, 219, 221
skulls, 64, 87, 127, 149, 158, 160
slash-and-burn agriculture, 238
slaveholders, 229, 230
"slow boat" model of Polynesian dispersal, 200–202
smallpox, 204, 205, 214, 215
Smith Woodward, Sir Arthur, 67
social adaptations, 118
social stratification, 210
social structure, 244
socioeconomic status, 217, 241
Solomon Islands, 198
Solutrean culture, 194, 196
South Africa, 68, 69, 89, 99, 105
South America, 198–202
South Asia, 202
South Asians, 195, 224
Southeast Asia, 202
Southeast Asians, 195
spatial ability, 118
spatial model of brow-ridge formation, 147
spears, 122, 142, 143
specialization, and bipedalism, 62
speciation, 43, 92
speech, 153–157
    see also language
Spencer, Herbert, 23
sperm production, 222
Sphinx of Giza, 208
"splitters", 92, 93, 98
stabilizing selection, 31
stepped pyramids, 209
Stone Age, 33, 203, 243
stone circles, 132–134

stone cores, 128, 131
stone tools,
    and animal bones, 132
    and *Australopithecus*, 89
    and bipedalism, 76, 77
    burned, 138
    and canine reduction, 65
    at Clovis, 193, 194
    and cognitive ability, 240
    cores, 128, 131
    cut marks, 140–142
    early, 65, 111, 118, 128
    at Flores, 179, 180
    and gut size, 119
    and *Homo* (genus), 5
    and *Homo floresiensis*, 182
    and *Homo habilis*, 110, 133
    and human origins, 63, 152
    and hunting, 139
    manufacture, 99
    and Neandertals, 152, 154
    at Olduvai Gorge, 132
    and species identification, 125–132
strategy, 152
Straus, William, Jr., 151
strength, 177
sub-haplogroups, 195
subpubic angle, 84
sub-Saharan Africans, 195
subspecies, 224
sunburn, 221
supraorbital torus, 144
"survival of the fittest", 22, 23, 24, 27
suspensory climbing, 75
Swartkrans site, 89
sweet potato, 202
symbolic behavior, 152, 154, 240

Tabun Cave, 138
tails, 53, 61
Taiwan, 200, 202
tanning, 219
taphonomy, 88, 99
tarsiers, 31

Taung child, 68–69, 89, 99, 116
technology, 2, 3, 100, 118, 158, 207, 243
teeth,
    and anagenesis, 40
    of apes, 47–48
    of *Ardipithecus*, 65, 78, 103, 105
    of *Australopithecus afarensis*, 103
    as clues to hominin ancestry, 46
    in *Gigantopithecus*, 58–60
    of *Homo*, 115, 126
    of *Homo habilis*, 110
    human, 9, 47–48, 60
    indicators of age, 83–84
    loss of, 204
    Neandertal, 149
    of *Paranthropus*, 94, 95, 96
    of *Proconsul*, 52, 53
    size, 40, 50, 54, 61–62, 65, 74, 82, 105–106, 110, 115, 126
    transition in human evolution, 103, 104
    wear patterns, 78
termite fishing, 121–123
tetrapods, 31
thalassemia, 239
theory, 7–11
thermoluminescence dating, 36
thymine, 163
tombs, 209
tools, 120–125
    Acheulian, 129–132, 137, 243
    use by *Australopithecus*, 85, 89, 128
    and bipedalism, 49, 76, 77, 80
    bone, 118
    bronze, 210
    use by chimpanzees, 63, 120–123
    chopping, 130
    flint, 142
    near fossils, 126
    hand axes, 130
    use by *Homo erectus*, 129–131
    human, 244
    iron, 210
    Levallois, 130, 131, 243
    Lomekwian, 128, 131
    Lower Paleolithic, 131
    Middle Paleolithic, 131
    use by Neandertals, 152, 154, 161
    Oldowan, 128–131, 133
    reliance on, 120
    as replacements for large canines, 48
    and selection for mental ability, 120
    and species identification, 125–132
    and tooth size, 65
    Upper Paleolithic, 131, 174
    waste flakes, 129
    at the Zhoukoudian site, 137
    see also stone tools
tooth marks, 141
traits, 30–31
    derived, 53
    human, 60–66
    Neandertal, 149
    recessive, 246
    unique, 62
transitional fossils, 101–106
Triassic period, 34
Trinkaus, Eric, 151
tuberculosis, 213, 216
Tyrannosaurus, 37

ultraviolet radiation, 16, 220, 221, 222
unique traits, 62
Upper Paleolithic, 131, 152, 174
upright walking, *see* bipedalism
uranium-lead dating, 36
Uruk, 210

vaccination, 214, 215
variation, 12–14, 27, 92, 223–227
Vietnam, 58
Vikings, 192
viruses, 8, 214, 216
visual cortex, 117
vitamin D, 222
vocal anatomy, 154

288 | Index

volcanic activity, 28, 36
Von Däniken, Erich, 207–209
von Koenigswald, Ralph, 58

Walker, Alan, 104
walking, *see* bipedalism
warfare, 205
Washoe, 125
waste disposal, 213, 215
waste flakes, 129
weapons, 48, 77, 87, 143, 152
Weinberg, Wilhelm, 249
wheat, 204
Wheeler, Peter, 119

White, Tim, 74
wildfires, 137
windbreaks, 132, 133
Wonderwerk Cave, 137
woodland environment, 76, 77, 79, 80
Woodward, Sir Arthur, 71
World Health Organization, 247
wrist bones, 179, 182

Yucatan Peninsula, 197
Yule, Udny, 248

Zhoukoudian site, 137, 138, 139, 141
zooarchaeology, 100